CELL CULTURE ENGINEERING

Technology and Applications

CELL CULTURE ENGINEERING

Technology and Applications

V. Radhakrishnan
Abhijith Anand

KNOWLEDGE
BAKERS

KNOWLEDGE BAKERS

ISBN: 978-93-49964-77-8

Published by : **Knowledge Bakers**
Pune, India
info@knowledgebakers.com

PREFACE

Cell culture engineering stands at the intersection of biological sciences, engineering, and biotechnology, playing a pivotal role in the development of pharmaceuticals, vaccines, tissue engineering, and bioprocessing technologies. The ability to cultivate cells in controlled environments has revolutionized medical research, providing critical insights into disease mechanisms, drug testing, and therapeutic development. As biotechnology continues to evolve, advancements in cell culture techniques are integral to improving the efficiency, scalability, and sustainability of cellular-based products and therapies.

This handbook, titled Cell Culture Engineering: Technology and Applications, offers a comprehensive exploration of the principles, methods, and innovations in cell culture technology. Designed for students, researchers, and professionals in the fields of biomedical engineering, biotechnology, and life sciences, this volume provides both theoretical foundations and practical knowledge essential for understanding and applying cell culture techniques in various applications.

The book delves into the key technologies used in cell culture engineering, including bioreactor design, media optimization, cell line development, and gene expression systems. It covers advanced topics such as 3D cell culture models, stem cell applications, cell-based therapies, and the role of automation and digital tools in enhancing cell culture processes. In addition, the book addresses critical challenges such as contamination control, scalability, and the optimization of bioprocesses for large-scale production.

Cell Culture Engineering: Technology and Applications also highlights the wide-ranging applications of cell culture in industries such as drug discovery, biotechnology, regenerative medicine, and vaccine production, offering case studies and real-world examples of successful implementations. The book aims to bridge the gap between the laboratory and industrial applications, equipping readers with the tools and knowledge to navigate the complex landscape of cell culture engineering.

As biotechnology continues to play a central role in addressing global health challenges, this handbook provides a valuable resource for anyone interested in advancing the science and technology of cell culture. It is our hope that Cell Culture Engineering: Technology and Applications will inspire new innovations and contribute to the continued development of this critical field in biotechnology.

V. Radhakrishnan

Abhijith Anand

CONTENTS

CHAPTER 1

AN OVERVIEW OF CELL CULTURE: CHALLENGES, PROSPECTS AND APPLICATIONS

1. INTRODUCTION: AN OVERVIEW

Cell culture, specifically Chicken embryo fibroblast cell culture, is a crucial laboratory technique extensively utilized in virology, vaccinology, molecular biology, microbiology, and biotechnology. This in vitro study replicates in vivo conditions for cell growth by controlling the environment, including medium, pH, CO2 level, temperature, humidity, and O2 flow. In recent decades, a small number of species have been studied using in vitro cell culture techniques, revealing important information about their biology. The chicken embryo is a valuable model for studying embryology, developmental biology, and producing medicinal proteins in transgenic chickens. It also contributes to research in virology, immunization progress, and vaccinology. Chicken embryo fibroblast cell culture significantly influences studies in veterinary vaccination and biochemical manufacturing. In addition, chicken stem cells offer an ideal chance for creating cell-based transgenic birds and serve as a potent cell source for manufacturing vaccines for poultry and human viral illnesses. Additionally, they provide insights into basic scientific concepts such as drug sensitivity testing, cell monitoring, and cytotoxicity testing. [1]. Many significant viral diseases have not been well studied at a cellular level, causing cell culture technology to remain in its early stages of development. This document outlines the fundamental procedures involved in establishing primary cell culture and explains the notion of cell culture technique. It aims to provide a practical guide for current and future researchers to utilize cell cultures effectively. This method has the ability to generate significant results and recommendations about CEF cell culture, cell metabolism, and response to various stresses and challenges. This work focuses on the fundamental concepts and essential techniques for primary cell culture, which can benefit both rookie and experienced researchers and laboratory professionals in this field, despite the majority of published papers focusing on CEF cell culture, BHK-21, and other cell lines. [2]. Physiological and physiochemical condition: Nature and impact of pH level, temperature, concentration of O2/CO2, and osmotic pressure (culture media) that can be altered the result of the study as well as their effects on the cell culture [3]. Cell culture technology has advanced significantly during the past century, resulting in the creation of several culture mediums. This chapter examines the basic principle, challenges, and current issues of cell culture, with a specific emphasis on chicken embryo fibroblast cell culture technology. This article is not intended to be exhaustive but rather aims to help researchers plan and make well-informed decisions for their research experiments or activities.

2. CELL CULTURE

Cell culture involves growing cells, fibroblasts, or live tissues in a specific media that supports their growth. Cell culture is the artificial growth of cells. CEF cell culture involves cultivating fibroblast cells derived from chicken embryos. Embryonated eggs are frequently utilized in the manufacturing of large quantities of antigens, vaccines, and other biochemical products. SPF eggs are sourced from SPF chicken flocks that are closely monitored for infectious agents and have not been vaccinated. Alternatively, they may come from healthy chicken flocks if necessary for specific purposes like producing certain inactivated vaccines and in accordance with regulatory approval. Laboratory personnel or researchers must select a specific route of inoculation for the SPF egg dependent on the study microbe (virus) being propagated [4].

Baby Hamster Kidney cells (BHK-21) are generally used in life science research work and the biopharmaceutical industry. Chinese hamster ovary (CHO) cells, baby hamster kidney (BHK) cells, mouse myeloma cells comprising NS0 and SP2/0, hybridomas, and human cell lines (HEK293, HT-1080) are the very commonly used mammalian cell lines at large scale [5]. Among these cell lines, BHK-21 has found applications at large scale in veterinary viral vaccines against foot and mouth disease (FMD) and rabies virus, as well as heterologous protein production (Factor VIII) [6]. The BHK-21 cell line was established in 1961 from the kidneys of 5 Syrian hamsters from litter number 21. Since this time, this cell line has been considered as a research facility standard for the development of countless viruses and the study of numerous biological processes. Baby hamster kidney (BHK) cells are one of many different vertebrate cell types used for the propagation of viruses by infection and transfection [7]. Generally obtained from BHK cells can vary widely if not obtained from the ATCC. For the production of animal products (vaccine) baby hamster kidney (BHK) cells are generally used, the most important of which is the production of a vaccine against FMD and rabies. Also, BHK was used in the production of recombinant proteins, such as blood coagulation factor VIII for the extraction of DNA from Pseudorabies virus and production of capture antigen enzyme-linked immunosorbent assay (ELISA) when diagnosing Japanese Encephalitis (JE) [8, 9, 10]. BHK-21 cells were grown and propagated in modified minimum essential media (MEM) by adding a set of proteins including Lactalbumin (2.50 g/l), yeast extract (1.00 g/l), peptone (2.50 g/l), New Zealand casein (1.00 g/l); Glutamine (0.50 g/l) and 2% sodium bicarbonate were also added to the culture medium.

3. BASIC STEPS OF CELL CULTURE TECHNIQUE

Cell culture, specifically CEF cell culture, is essential laboratory activity. Lab personnel or researchers must maintain a sequential process during cell cultivation. Cell culture should be conducted smoothly by adhering to the following steps. SPF eggs were obtained from reputable sources with proper documentation for CEF cell culture.

Then incubate for the recommended date. Afterward-

- Personal hygiene and safety management
- Sterilization and disinfection
- Arrangement of instrument and appliances as well as power supply
- Biosafety cabinet (BSC) management
- Management of essential ingredients and chemicals like Media, FBS, Trypsin, Phosphate Buffer Saline (PBS), Tincture of iodine, Antibiotics, etc.
- Management of Pipettes, Burette, Tips, Waste disposal box, Aspirator, Tissue
- Chopping, washing embryo, and filtering (for CEF cell culture)
- Centrifugation
- Cell suspension and stock management
- Cell count and splitting of the cell
- CCF, Media, and incubator management
- Cell line development
- Cell observation, harvest
- Cell count, subculture, infection, and bulk antigen production

4. FUNDAMENTAL PARTS OF CELL CULTURE TECHNOLOGY

4.1 Culture medium

Culture medium is a composition of nutrients and selected buffer that helps to grow an organism naturally. Media can be designed based on variety of cell, types of cells because it is necessary for cell survival, proliferations and growth. The influence of cell culture technology creates inevitable progress in molecular biology research. This technique widely utilized in different fields like the assessment of toxicity and efficacy of new drugs, development of various biopharmaceutical products and vaccines, and used in reproductive technology. No one probably would argue against the claim that a culture medium is the foremost essential measure in cell culture technique. Selection of suitable media for research goal is essential. Sometimes researchers should modify a properties or composition of medium in order to their experiment. There are mainly two types of media used by researcher such as natural media and synthetic media [3, 11, 12]. Natural media: The media that comprising of natural biological substances, e.g. embryo extract, plasma, and serum. Coagulant or clots i.e. plasma which is separated from heparinized blood, serum, and fibrinogen. Tissue extracts and Extracts of chicken embryos, liver, and spleen and bone marrow extract considered as a natural cell culture media. Biological fluids such as Plasma, serum, lymph, amniotic fluid, and pleural fluid also used as culture media. Synthetic media: The media comprising with basal medium and supplements, like-serum, growth factors, and different hormones. On the basis of type of supplements synthetic media can be classified into a few bunches such as: (1) Serum-containing media: In this media as a supplement bovine, human, equine, or other serum is used. This media causes less reproducibility and microbial contamination due to unknown and unclear media composition. (2) Serum-free media: This media has a high reproducibility due to presence of the supplements of bovine serum albumin or α-or β-globulin (crude protein fractions).

(3) Xeno-free Media: Prepared based on human-source components i.e. human serum albumin, are commonly used as supplements but there are no animal components used as supplements. (4) Protein-free media: Prepared based on undefined components like- peptide fractions (protein hydrolysates) which are generally used as supplements, and (5) Chemically defined media: Undefined components are used as supplement. Crude protein fractions, tissue extracts, hydrolysates are not good supplements, but highly purified components (recombinant proteins) are appropriate supplements. Followings are some brand name of culture media commonly used in the laboratory: Eagle's Minimum Essential Medium (EMEM) a simpler basal medium (BME) for culturing mouse L cells and HeLa cells. Dulbecco's Modified Eagle's Medium (DMEM) is a basal media (BME). Iscove's Modified Dulbecco's Medium (IMDM) used for growth of lymphocytes and hybridomas. Hybri-Care Medium is a combination and modification of DMEM considered for the propagation of hybridomas and other fastidious cell lines. Ham's Nutrient Mixtures used for Chinese hamster ovary (CHO) cells. Kaighn's modification of Ham's F-12 (Ham's F-12 K) was designed to develop and differentiation of primary cells with or without serum. DMEM/F12 Medium is a 1:1 mixture of Dulbecco's modified EMEM and Ham's F-12 for the growth of a wide range of cell types in both serum and serum-free formulations. Leibovitz's L-15 Medium for biopsy samples without CO2.

4.2 pH value

pH plays a vital role in cell culture. The cell growth rate is decline associated with Fluctuations in pH level. That's why routine monitoring is essential. For the cells is 7 the optimal pH, and decline or increase in pH can hinder the growth of cells. More decrease in pH level (usually in between 6.0–6.5), can stop the growth rate of the cells, and cells are start losing viability at low pH level. That's why pH level should be maintained and monitored carefully for individual cell line. If pH level fall rate is less than 0.1 units/day, that indicates the cell condition is good and no need to hurry to change the culture medium immediately. If it is 0.4 units/day (pH drop rate), that indicates the culture medium need to be changed quickly [13]. Alkali (like- NaOH, KOH) or acid (HCl) solution helps to control pH level in culture medium. Besides, NaHCO3 (sodium bicarbonate) or natural buffer solution, and addition (need base) of CO2 gas to the bioreactor also helps to maintain optimum pH level in culture medium. Generally, pH electrode (silver chloride electrochemical-type) used within the bioreactor [14]. For the proliferation of cell in culture medium an optimum, stable as well as balanced pH is essential. Depending upon cell type and culturing process pH level may vary and specific. Generally, 5–10% CO2 required to grown cells using buffered media that contain NaHCO3 (sodium bicarbonate) and where maintained the range of pH 7.2 to 7.4 [15]. CO2 incubator, optimum pH level, ideal temperature, optimum moisture condition, sterile and clean working environment are essential to maintain and complete an experiment. In the cell culture medium, the carbonate buffer helps to hold constant pH and take parts in releasing CO2 gas in the CO2 incubator. Color of the culture medium depending and changing with the pH level of culture medium. Color indicates altering the medium and CO2 levels [3]. Commonly, 4–10% of CO2 is practiced in the cell culture technique. By maintaining HCO3- concentration and CO2 tension in culture medium one can easily achieved optimum pH and osmolality [16]. By observing the color of the media can easily identify the pH condition like- Phenol red to yellow/ orange color indicates too acidic where pH 6.8 (bellow), Red to pink color indicates pH above 7.0 to 7.7 which is normal, and bright fuchsia color indicates pH 8.0 to 8.2 (too alkaline).

4.3 Effect of temperature

Generally, cell culture need 37°C for incubation called control temperature. Proliferation and multiplication of cells are significantly decreased at more than 40°C, like 41°C or 42°C temperature, and increased temperature may also cause high apoptotic rate of CEF cells. Cell viability, apoptosis, proliferation, and oxidative status of cells in culture medium can be altered with high temperature. ROS (reactive oxygen species) formation increased with increasing Temperature (Proportionally) [17]. During the transportation must be care full about temperature. Besides regular or routine inspection is recommended for better results. In cell culture technique one of the most challenging issues is to grow cell, that's why an ideal temperature play a vital role in cell culture along with good supplementation of nutrients. For the cell division the optimum temperature should be vary on cell type that assist to maintain growth rate. Generally, at optimal temperature metabolic function of cell is optimum as well as good that helps to increase their size, and proliferation rate [18]. Temperature requirement varies based on cell type like- S. frugiperda (sf-21, Trichophusia ni (Tn-5), S. frugiperda (sf-9) (insect) need 27°C, Chinese Hamster Ovary (CHO) (Mammalian)-need 36–37°C, Human embryonic kidney (HEK- 293 T) (Human) need 37°C, HF 205 and HF 210 (plant) need 27°C.

4.4 Incubation

A device in which microbial culture or cell culture is grown and maintained with customized temperature, humidity (relative humidity 95%), oxygen, CO2 level, and other conditions. For virus cultivation, cell culture as well as cell infection, vaccine development incubation is a very much crucial and fundamental issue. The incubator is an essential instrument in cell biology research, Microbiology, Biotechnology, Molecular biology research. On the other hand, an egg incubator is one of the most important for embryonated egg production in the laboratory and large scale. BOD (Biological Oxygen Demand) incubator is popular in this regard. There are so many necessary points always bear in mind to maintain good incubation. Firstly, cleaning inside, cabinets, outside and handle that helpful to eradicate cross-contamination, also helps to hold a good quality of cell, media, SPF (Specific Pathogen Free) eggs, and other chemicals or ingredients. Logbook maintenance is another vital issue to maintain high-quality research work. Through logbooks, laboratory personnel can easily identify any issue related to the incubator. A dedicated power supply essential to maintain cell quality, growth, metabolism as well as important to maintain cell physiology. CO2 level and the supply are very much essential to maintain moisture and pH (normal range is 7.2 to 7.4). Generally, 5% CO2 with 37°C is used to maintain the cell. Eventually, inventory management will helpful for GMP (Good Manufacturing Practice) and proper cell culture as well as laboratory work.

4.5 Cryopreservation

Cryopreservation is the process of preserving living cells by storing them intact in liquid nitrogen at cryogenic temperatures. Conversely, in cryopreservation, utilizing low temperatures aids in maintaining the structural integrity of living cells. Cryopreservation technique preserves living cells for extended periods, often for years. Cells are protected from harm during long-term preservation and chemical reactivity due to the cessation of their regular metabolic activity produced by freezing temperatures. Handle cells delicately to prevent damage and stress during the freeze-thaw process. Optimal cryopreservation of cells relies on proper freezing and thawing methods. An effective cryopreservation technique is determined by assessing the recovery rate of frozen cells at low temperatures, the percentage of viable cells, and the rate of cells that maintain normal function after thawing. [19]. Basically, for cryopreservation harvest the cells in exponential growth. Then gentle centrifugation done at 125 × g for 10 minutes. After that check and scree, the media. The procedure starts with taking freezing medium (GM-which warm at room temperature for 30 minutes) that containing a cryoprotectant such as Dimethyl sulphoxide (DMSO) (e.g. 5–10% v/v DMSO), fetal bovine serum (FBS) (10–20% v/v) and at high cell density (1–5 × 106/ml) and sometimes added knock-out serum replacer (KoSR; 20% v/v final), bovine serum albumin (5% final) or human serum albumin. From some recent research [20, 21, 22] it is said that freezing rate has great impact on viability of cryopreserved cells. It is suggested that cells be slowly cooled (like 2°C, –20°C, –80°C for at least 24 hours and finally preserved in liquid nitrogen at –196°C) that gives better surviving rate of cell in the cryopreservation process. Record logs must maintain during all the steps. From the final cryovial (that contain cell at –196°C) remove one vial and restore the cells in culture medium to determine cell viability and sterility. Recovery rate of cryopreserved cells depends on the types of cultured cell. Some cell needs several days, some shows low viability on the day of culture, in some cases cell produce debris, some cells are shows normal viability after 24 hours' post-thaw. Before retrieving of cryopreserved cells clean the biosafety cabinet, prepare the CCF, media, FBS and arrange all the instruments and appliances (sterile). Then wash the cryovial with 70% ethanol and place it in a water bath for 2 to 5 min at 37°C to melt. Transfer the thawed cell in a tube and gentle centrifugation (10 minutes at 125 × g) needs to discard (supernatant) cryoprotectant in the meanwhile collect the cell pellet and suspend the cells in 1 mL or 2 mL of complete growth medium (GM) then proceed for cell count and subculture in new CCF for 24 hours' observation. In the process of cryopreservation, significant rate of cell survival and maintenance of cell integrity (structural and morphological) can be achieved by using cryoprotective agents (CPAs). Excipient is an ingredient added intentionally to the drug substance which should not have pharmacological properties in the quantity used. Commercially available CPAs namely Dimethyl sulphoxide (DMSO) is commonly used as CPA. Factors behind the success of cell survival [23] are: (a) Type and concentration of cryoprotectants (an additive, such as glycerol or dimethyl sulphoxide, that can protect cells against freezing injury). (b) Cell density in cryopreservation solution at the time of freezing. (c) Cooling and thawing rates of cell suspension. (d) Dilution rate of thawed cell suspension. The main advantages [24] of cryopreservation are easily found original cell lines from the safety stocks, preserve the cells for year after year and lastly smoothly perform continuous research or experiments.

4.6 Thawing

Correctly thawing of cells is crucial to recover quickly, yielding the highest viability and functionality possible. Some cryoprotectants (e.g. DMSO) has toxic effect on cells, due to the possibility of toxicity cells should be thawed rapidly and not allowed to remain in the freezing medium no longer than required time. Firstly, retrieve the cryovial containing the frozen cells from liquid nitrogen (-80°C or - 196°C freezer) and immediately place it into a 37°C water bath (for 1 to 2 min) or place immediately in a pre-equilibrated thermo-conductive rack or tube module resting on dry ice to minimize cell warming/thawing. Rapidly thaw the cells (< 1 minute) by carefully swirling the cryovial in the 37°C water bath up to there is a little bit of ice left within the cryovial. Then place the cryovial into a BSC. Gently wipe the outside of the cryovial with 70% IPA (Isopropyl Alcohol) prior to open the cryovial screw. Carefully add (dropwise) required amount of pre-warmed GM into the tube (centrifuge) that containing recently thawed cells. Place the cell suspension in centrifuge machine for centrifugation and set 200 × g for 5–10 minutes (it may vary based on cell type). Inspect the transparency of supernatant and visibility of a pellet at the bottom after completing the centrifugation. Discard the supernatant aseptically without breaking the pellet. Softly resuspend the cells by gently pipetting with GM and prepare required concentration, then transfer the cell suspension into the CCF, and place it in the suggested culture environment. Inspect the cells using an inverted microscope for morphology. Examine an aliquot of cells for the ability to exclude trypan blue. If cells pass both inspections, they are ready for culture [25].

4.7 Passaging of cells

Subculture of cell commonly known as passaging of cells and the ratio of subculture is 1:2. The main concept of passaging: cells are split into half in each subculture. Continuous cell lines should will be passaged with higher split ratio due to their higher replication rate. Usually, the number of times the cells have been subcultured into a new CCF known as passage number. In the case of diploid cell cultures, the number cell passage is partially equal to the number of population doubling level (PDL) since the culture was begun. On the other hand, PDL of continuous cell lines is not fixed like diploid cell culture. Mostly the PDL is an estimation or prediction. PDL may ups and down with cell stress and cell death (due to necrosis, apoptosis). Loss of proliferation capacity of cell, contamination of culture medium may also responsible. A common formula for the calculation of population doubling level: PDL = 3.32 (log Xe – log Xb) + S; where Xb is the cell number at the beginning of the incubation time, Xe is the cell number at the end of the incubation time, S is the starting PDL. Another common formula used to calculate the population doubling level is [26]: Log (N/N) x 3.33; Where N is the number of cells in the culture vessel at the end of a certain time interval, and N is the original number of cells plated in the vessel, Population doubling and passage number are often mixed up or thought to mean the same thing. The passage number describes the number of times that a culture has been subcultured The population doubling time is important to know when the number will be double and passaging will be required. According to ATCC [24] the population doubling time calculate with the formula: DT = T ln2/ln(Xe/Xb); where T is the incubation time in any units, Xb is the cell number at the beginning of the incubation time, Xe is the cell number at the end of the incubation time.

4.8 Trypsinization

Cryopreservation is the process of preserving living cells by storing them intact in liquid nitrogen at cryogenic temperatures. Conversely, in cryopreservation, utilizing low temperatures aids in maintaining the structural integrity of living cells. Cryopreservation technique preserves living cells for extended periods, often for years. Cells are protected from harm during long-term preservation and chemical reactivity due to the cessation of their regular metabolic activity produced by freezing temperatures. Handle cells delicately to prevent damage and stress during the freeze-thaw process. Optimal cryopreservation of cells relies on suitable freezing and thawing procedures. An effective cryopreservation technique is determined by assessing the recovery rate of frozen cells at low temperatures, the percentage of viable cells, and the rate of cells that maintain normal function after thawing [27]. Commonly 0.05% trypsin used in laboratory work. Sometimes trypsinization causes cell damage and sometimes may not effective for some cells thus other dissociation agents (enzyme) are recommended for the dissociation of cells. Warm and cold trypsinization are the two common approaches. An extensively used method is warm trypsinization. In which cells are washed with basal salt solution and then add warm trypsin (37°C) adequately and stirred properly. The supernatant dissociated, the cells are dispersing in the medium. In the case of cold trypsinization, cellular damage is reduced, resulting in a high yield of viable cells also improved survival rate. For cold trypsinization cells are maintained in ice after washing with media or salt then treated with cold trypsin for 6–24 h. After that remove and discard the trypsin and incubate the CCF at 37°C (for 20–30 min). Dispersion of cells may start and fully dispersed cells counted using hemocytometer then dispersing in a medium for further use. The easiest way of trypsinization is a). Discard the media from CCF b). Wash the cell surface with PBS (4 ml for 75 cm2 CCF) c). Take trypsin (Room temperature) d). Rinse monolayer of cell with trypsin–EDTA (2 to 4 ml for 75 cm2 CCF) e). Stay for 2 minutes then discard trypsin and Incubate the CCF at 37°C for 5 minutes f). Tapping the CCF and collect the cells by scrapper stored in a tube g). Spin down the cells, resuspended by adding the growth medium or fresh medium, and Count the cells h). Split into a new flask and Incubate at 37°C.

4.9 CCF measurement

CCF denotes a Cell Culture Flask. It is also known as a tissue culture flask. CCF is important for culturing of cells, transportation of cells, and media. There are a lot of different volumes of CCF used for research work. Commonly, used flask volume are 25 cm2, 75 cm2, 175 cm2, 225 cm2, 300 cm2. Commonly 75 cm2 CCF preferable for laboratory work. Cell contamination generally appeared during cell culture laboratory. Proper knowledge of CCF handling can minimize cross-contamination, which improves the quality and physiology of the cell. Rough handling of CCF during media transfer, passaging, scrapping will be responsible for the different vital issues. Mycoplasma contamination is one of them. Prompt and Improper pipetting during cell harvest and split from CCF may cause stressful conditions on cells and resulting in cell death. Scrapping of the cell for subculture or infection or cell count gently handles the CCF, corkscrew. Mild flame spark on corkscrew (CCF) by flame gun or gas burner helpful to safe the cell and flask environment. Covering the cork with Paraflim very much essential to save the cell. IPA (Isopropyl Alcohol) spray must be done before and after handling of CCF. After application of IPA, then CCF, media, FBS (Fetal Bovine Serum), tips, flame gun, trypsin other materials, and appliances allow entering into BCS (Biosafety Cabinet) for further processing. Some points should be bear in mind regarding CCF such as, is CCF allow pipettes, tissue scrapper, tissue spatula properly? Is CCF has marked on both sides? Is there any leakage? Is there any crake on cork? Is the bag of CCF tightly pack? Is the CCF clean (inside)? These points might be helpful for cell culture.

4.10 Cell suspension

Cryopreservation is the process of preserving living cells by storing them intact in liquid nitrogen at cryogenic temperatures. Conversely, in cryopreservation, utilizing low temperatures aids in maintaining the structural integrity of living cells. Cryopreservation technique preserves living cells for extended periods, often for years. Cells are protected from harm during long-term preservation and chemical reactivity due to the cessation of their regular metabolic activity produced by freezing temperatures. Handle cells delicately to prevent damage and stress during the freeze-thaw process. Optimal cryopreservation of cells relies on suitable freezing and thawing procedures. An effective cryopreservation technique is determined by assessing the recovery rate of frozen cells at low temperatures, the percentage of viable cells, and the rate of cells that maintain normal function after thawing.

4.11 Cell count and cell infection

Cell counting was performed using a hemocytometer (Neubauer improved counting chamber, Precicolor HBG, Germany) or MacMaster slide and Trypan blue exclusion every 24 h (1,1 mixture of 0.2% Trypan blue in normal saline solution and sample). After placing the stock cell suspension on the hemocytometer and place a coverslip on it. Count cell of 4 (16 × 4) site, then the average of 4 sites of hemocytometer and count the total cell as = Average (number of the cell) × 10,000 × 2. Cell culture flask such as 25 cm2 contain 5 to 10 ml culture media, 75 cm2 contain 10 to 30 ml culture media, and 175 cm2 contain 40 to 150 ml culture media. One (1) cm2 need 90,000 cells likewise 75 cm2 need 75 × 90000/5 = 13,50,000 cells minimum. Viable cells are considered as unstained ones and while stained cells are considered as dead under the inverted microscope. Cell counts are important for monitoring cell health and proliferation rate, assessing immortalization or transformation, seeding cells for subsequent experiments, transfection or infection, and preparing for cell-based assays. Cell counts must be accurate, consistent, and fast, particularly for quantitative measurements of cellular responses. Cell infection is required for virus propagation, bulk antigen production. MOI rate is very much essential in cell infection. There are three types of MOI commonly used in the laboratory such as 1 MOI, 0.1 MOI, 0.001 MOI where 1 MOI means one (1) virus can infect one (1) cell. 0.1 MOI denotes 10 cells infected by one (1) virus and one virus can infect 100 cells in 0.001 MOI. Generally, practice 0.1 MOI means one virus is enough to infect 10 cells.

4.12 PFU (PFU), ELD50, EID50, MOI, CCID50, TCID50

Plaque forming units (pfu) is an assessing of the total number of infectious virus particles. It is ascertained by a plaque-forming assay. In the field of virology study, a plaque-forming unit (PFU) is a measurement of the number of particles capable of forming plaques per unit volume i.e. virus particles. It is a functional measurement rather than a measurement of the absolute quantity of particles: viral particles that are defective or which fail to infect their target cell will not produce a plaque and thus will not be counted. For instance, a solution of virus with a concentration of 1,000 PFU/μl indicates that 1 μl of the solution contains enough virus particles to produce 1000 infectious plaques in a monolayer cell, but no inference can be made about the relationship of pfu to the number of virus particles.

- **ELD50-** Embryo Lethal Dosage. ELD50 unit is the amount of virus that will kill 50 percent of inoculated eggs.
- **EID50-** Embryo Infective Dosage. EID50 unit is the amount of virus that will infect 50 percent of inoculated eggs.

Multiplicity of infection (MOI) is the average number of virus particles infecting each cell. MOI is related to pfu by the following formula: Multiplicity of infection (moi) = Plaque forming units (pfu) of virus used for infection/number of cells. For example, if 2x106 cells is infected by 50 ml of the virus with a titer of 108 pfu/ml. The moi will be 0.05*108/2*106 = 2.5. The fraction of cells that are not infected is P(0) = 1 - e–moi. To ensure 99% of cells are infected requires moi > 4.6. Assume the conditions used for plaque assay and TCID assay do not alter the expression of infectious virus. TCID50/ml and pfu/ml are related by pfu/ml = 0.7 * TCID50. As a working estimate, one can use pfu/ml = 0.5 * TCID50 [28].

- **CCID50:** Cell culture infectious dose which will infect 50% of the cell.
- **TCID50** is the tissue culture infectious dose that will infect 50% if the cell monolayers are challenged with the defined inoculum. Two methods commonly used to calculate TCID50 (can also be used to calculate other types of 50% endpoint such EC50, IC50, and LD50) are a). Spearman-Karber [29] b). Reed-Muench method.

4.13 Plaque assay

Plaque-based assays are the standard method used to determine virus concentration in terms of infectious dose. Viral plaque assays determine the number of plaque-forming units (pfu) in a virus sample, which is one measure of virus quantity. This assay is based on a microbiological method conducted in Petri dishes or multi-well plates like 6 well or 24 well etc. Specifically, a confluent monolayer of host cells is infected with the virus at varying dilutions and covered with a semi-solid medium, such as agar or carboxymethyl cellulose, to prevent the virus infection from spreading indiscriminately. A viral plaque is formed when a virus infects a cell within the fixed cell monolayer [30]. Virus quantification involves counting the number of viruses in a specific volume to determine the virus concentration. In research and development (R&D) based commercial and academic laboratories, the production of viral vaccines, recombinant proteins using viral vectors, viral antigens and clone screening, multiplicity of infection (MOI) optimization, and adaptation of methods to cell culture all require virus quantification. For quantification of virus incubated after infection at 37°C in a 5% CO2 incubator at 6, 12, 24, 36, 48, 60, and 72 h post-inoculation (hpi) based on the requirement to visualize plaques in wells [31, 32]. To quantify virus there are a lot of other methods used such as Focus forming assay (FFA), Endpoint dilution assay, Protein assays, Hemagglutination assay (HA), Bicinchoninic acid assay, Single radial immunodiffusion assay, Transmission electron microscopy (TEM).

4.14 Cell harvest

Knowledge of splitting, media, trypsinization, cell handling is essential to harvest cells. Firstly, remove and discard the media from CCF. For 75 cm2 CCF, mild PBS wash is required before the application of warm trypsin or trypsinization process. Add trypsin (2–4 ml/75 cm2) then incubate as like trypsinization process. When detached cells appear then add 2–5 ml growth media (GM) to inactivate trypsin. Gently pipette to disperse the medium to ensure recovery of >95% of cells. Sometimes commercial trypsin inhibitor is added. Carefully centrifuge the collected cell suspension at 300–1000 X g for 5–10 min. Discard the supernatant and add GM at the required amount for the preparation of cell count. Split the cells after counting or go for further processes that need. There is a lot of problems that may be appeared like detachment difficulty of cells from culture flasks, cell adherence difficulty, insufficient attachment of cells, low viability of cells, clumping after detachment, damage of cell membrane, and cell death. The possible solutions to the above problems will be careful during the following such as a). Check the quality, date, and concentration of trypsin before use b). Be careful during antibiotic application if any c). Splitting, media replacement required before harvesting if cells are in stress d). Avoid vigorous pipetting and long centrifugation.

4.15 Antibiotics

Routine cell and tissue culture according to good cell culture practice (GCCP) [33] should not require the use of antibiotics as they can never be relied on as a substitute for effective aseptic techniques. However, its use is still widespread e.g., OECD TG 432 [34] due to established routine procedures in many laboratories. Antibiotics are agents that may arrest or disrupt fundamental aspects of cell biology, and, while they are effective against prokaryotic cells (i.e. Bacteria), they are also capable of causing toxic effects in animal cells. Not surprisingly, antifungal agents, being directed at higher order, eukaryotic microorganisms, are likely to be more toxic to animal cell cultures. In addition, antibiotics often make it more difficult to detect microbial contamination. These obvious contraindications, the use of antibiotics in cell and tissue culture should be focused in two areas: a) Protection of materials at high risk of contamination such as tissues, organs, and primary cultures in cases where sterility cannot be guaranteed, and b) The positive selection of recombinant cell clones based on the expression of antibiotic resistance genes [33]. If antibiotics are needed, a justification for the use of antibiotics in the procedure is suggested.

4.16 Laboratory management

Good laboratory setup is essential for cell culture as well as good laboratory practice is also essential for better and smooth work. The following discussion and points are very much important for a cell culture laboratory and are also supported by Maneesha et al. [35] and Coecke et al. [36].

Location and ideal layout with the purpose-built facility is the first priority. Room data sheet (RDS) in which available all the data of every facility, specification of the room define its location, a number of doors, windows, pass box, ventilator, light, air conditioner, fire alarm like everything present on that room permanently. Specification of instruments must be available at the working area that helps the laboratory personnel to operate it. Standard operating procedure (SOP) helps to do research/ laboratory work in a defined way to get a better outcome. Define the area of a room with a specific class and biosafety level by the standard of ISO, GMP. Define or specify works of laboratory personnel with defined working areas according to CDC, NIH-USA also beneficial for a good outcome. There are four biosafety levels based on hazard or pathogenicity or virulence of microorganism and toxicity of agents. The basic biosafety level known as biosafety level 1 (BSL-1). In this level generally very common research work done with easy protection. Normally in BSL-1 working with those organisms or agents which are not harmful to healthy person. In biosafety level 2 (BSL-2) working with those organisms or agents which is known as moderate-risk agents, and known as a potential threat to human resulting produce disease (by ingestion or through percutaneous or mucous membrane exposure) of varying severity. Generally, cell culture should be performed at BSL-2 laboratory. Sometimes the biosafety level depends on the type of cell line and working style. In biosafety level-3 (BSL-3) working with that agents which have the capability to transmit through air (aerosol transmission). BSL-3 agents may be indigenous or exotic, having potential threat to human, may cause serious health issue, and may be potentially lethal. Biosafety level-4 (BSL-4) deals with exotic agents that create life-threatening disease of an individual through infectious aerosols and for which no treatment is available. These organisms or exotic agents are restricted to high containment laboratories. The easily accessible facility of the emergency shower should be available in the laboratory area. Access control in laboratories should be helpful to maintain unwanted occurrences. A separate logbook is very much helpful for liquid nitrogen and CO2 management. Ventilation and pressure control like negative and positive pressure controlled area must be defined, HEPA filtered providing positive pressure to clean areas, is recommended where space and resources allow. Electricity room (uninterrupted power supply (UPS) units should be provided for essential equipment (class II cabinets, incubators, air filtration) and to allow cell culture procedures to be completed. Accountability for all the staff will provide a smooth working environment. Documentation, Training (Fumigation, 5 s, SOP, etc.) and Monitoring of staffs, Emergency service provider contract with the third party are essential for good laboratory management. Before receive and entering all the reagents, chemicals, and supplier's documents must be checked by maintaining a logbook. For the management of inventory and documents should be established hard copy or electronic form that stored the information of materials, cells, suppliers, overall all the possible information. Staff safety is a vital issue. The primary concerns regarding safe management of liquid nitrogen storage are frostbite burns from skin contact with liquid nitrogen and asphyxiation due to exposure to low oxygen levels when nitrogen gas is released from vessels. Finally, all service personnel entering the laboratory should receive instruction in special laboratory hazards and any necessary procedures for working in clean areas e.g. gowning, hand disinfection. All the person needs to use a separate biohazard bag/ bin to ensure safe hazard management. Like- Infectious non-sharp waste (incineration/deep burial)-Yellow bag; Plastics and sharps (chemical treatment/autoclaving/shredding/microwaving)- Blue bag; Infectious non-sharp waste (chemical treatment/autoclaving/ microwaving)-Red bag; Incineration ash and solid chemical wastes (secure landfill)- Black bag.

4.17 Challenges during passage of cell

Cell cultivation requires laboratory personnel and researchers to regularly monitor the cells. Cell division is influenced by factors such as the cell's doubling rate, cell type, pH level, culture media, and various other cell culture-related factors. Laboratory workers encounter numerous challenging situations. Common issues in the laboratory that can significantly impact operations include misidentification of cell lines, contamination of cultures and media, rough handling, poor cell growth, poor cell attachment, improper trypsinization during harvesting, incorrect cell counting, improper split ratios during passages, cell clumping, and cell death. Incubation duration, temperature, and CO2 concentration significantly affect both cell culture and subculture.

4.18 Maintenance of cell

The routine follow-up of cell morphology is necessary. To maintain a good cell line routinely change of the medium is essential for the both proliferating or non-proliferating cells. The culture medium should be changed repeatedly in the case of proliferating cells compared to the non-proliferating one. The rate of cell growth, cell morphology and metabolism of cell indicates the urgency and time interval of medium change. For example, HeLa cells are rapidly growing transformed cells, in the case of HeLa cell the culture medium should be changed twice within 7 days, whereas for slowly growing non-transformed cells (like IMR-90 cells) the culture medium may be changed once in a week. Continuous cell lines, Chicken embryo fibroblast cell (CEFC), Embryonic cells and transformed cells develop quickly that's why these cells need rapid sub-culture and altering the culture medium. While normal cells are grow slowly. Generally altering the medium depends on pH level. Immediately change whole medium when the pH level appeared 7.0, cells are stop proliferating at pH 6.5, and the cells may lose their durability and viability when the pH level drop in between 6.5 to 6.0. The drop rate of pH is commonly estimated for each and individual cell line with a selected culture medium. If the drop rate of pH is less than 0.1 units/day, that indicates no harm and no need to hurry to change the culture medium immediately. When the drop rate of pH is 0.4 units/day, that indicates the culture medium need to be changed immediately [13]. A laboratory person or a researcher can easily maintain cells by maintaining SOP of cell handling and culture procedure, by counting passaging time because 10–30% density of cell is standard but at 80–90% density cell should be split as well as cell count may be helpful in this regard. Must pay attention to media quality, color, clarity, foul smell results from infection of the cell. Cell health and cell concentration, appearance observed regular interval by bright field microscope with 20–60x magnification may help to maintain cells and eradicate clumping, detachment, apoptosis. QC papers, COA, and Cell transportation SOP are essential for cell maintenance. A logbook entry is necessary to record daily activities such as pH level, medium condition, temperature, CO2 level, information of assigned personnel, cell condition, passage number, and other critical details to maintain cells for small or large-scale work.

4.19 Mycoplasma: an issue

Mycoplasma contamination is a significant and prevalent issue in cell culture. Mycoplasma is frequently transmitted between cultures and laboratories. Mycoplasma contamination can compromise an entire research project if data is collected from cells or cultures that are infected with mycoplasma. Mycoplasma is a type of biological contaminant in laboratories that can spread quickly and have a harmful effect on cells. They are difficult to identify and can seriously influence cell lines. Although mycoplasmas are minute organisms similar to bacteria, they possess unique traits that set them apart. Mycoplasmas can thrive and reproduce at high concentrations without exhibiting any visible symptoms. They are extremely detrimental to any cell culture. Mycoplasmas can readily modify the metabolism and shape of host cells, leading to chromosomal abnormalities and cell damage that trigger cytopathic consequences. Testing for Mycoplasma should be conducted monthly in laboratory and research settings. Utilizing DAPI staining and PCR are beneficial testing procedures, however it is advisable to also use a commercially available mycoplasma kit for detection. [37]. A routine screening process might be helpful to eradicate mycoplasma contamination from the lab. The following points are essential to prevent mycoplasma issues [38]. a). Wearing personal protective equipment (PPE) during cell culture that includes a dedicated, clean lab coat and gloves b). Checking the COA, QC pass of cells' origins c). Ensure proper sterilization d). Always clean the working area e). To avoid cross-contamination work with only one cell at a time f). Always ensure covering the media bottles and CCF. Do not use the hood for storage and work always within biosafety cabinet g). Should be cautious in the use of antibiotics because it is reported that antibiotics have no impact on mycoplasmas h). Maintain logs for record-keeping that help to identify possible contamination sources if needed i). SOP develop for routine mycoplasma screening.

4.20 Bulk antigen and vaccine candidate development

Bulk antigen manufacturing refers to the large-scale synthesis of viruses. Establishing a suitable cell culture, transferring cells, infecting cells with a specific virus at the desired Multiplicity of Infection (MOI), collecting the virus, filtering it, doing quality control tests, and obtaining a Certificate of Analysis (COA) provide the ultimate verification of the virus and the procedure. Following formulation, dosing should be carried out using bulk antigen to aid in the development of the proposed vaccine candidate according to the reference manual and guidelines.

5. CELL CULTURE PROCEDURE

CEF cell culture is widely used by researchers in the biopharmaceutical industry and veterinary vaccine production. The following steps should be maintaining chronologically to develop primary cell culture from the chicken embryo.

- SPF eggs are incubated and collect the embryonated egg at 8–11 days (need base).
- Cleaning the outside of egg by tincture of iodine, Cracking the egg into BSC.
- Collect the embryo by forceps, place it into a sterile Petri dish, and chopping the body parts (embryo) with scissors.
- Cut off head, wings, legs, remove the visceral parts and wash (PBS) rest of the body in Petri dish until clean the blood.
- Chopping the clean body with scissors and gently pipette and aspirate by syringe.
- Collect the suspension into a sterile falcon tube and perform mild centrifugation at 300 rpm for 5 min.
- Add 0.25% trypsin EDTA to suspend the pellet by gentle pipetting (Recommendation: for 12 embryos add 10 ml trypsin and 10 ml PBS for suspension).
- Centrifuge the suspension at 600 to 1000 rpm for 5 to 10 min (at 37°C).
- Collect the supernatant into a new sterile falcon tube and filter it with double layer sterile gauze and collect it into a new tube and add 10 to 15 ml GM into it for 12 embryos. Generally, add GM two times (1:2) of collected cell/fluid.
- Wash it by centrifugation at 10,000 rpm, 25°C for 10 min, and discard the supernatant. Then add GM to reconstitute and gently pipette. Repeat this step 2 times and collect the pellet (Suspected that 12 embryos produce 2 ml pellet).
- Add 13 ml GM with the collected pellet (2 ml) and gently mix 15 ml cell suspension consider as a stock cell suspension.
- Count the cell based on the stock with the help of a hemocytometer then split the incorporation of the cell

with growth media that contain 10% FBS based on CCF measurement.

- Observe the cell distribution within the CCF under an inverted microscope.
- Place the CCF into a CO2 incubator at 5% level, pH 7.2 to 7.4, 37°C for overnight. Then observe the cell morphology under a microscope.

6. APPLICATION OF CELL CULTURE

Nowadays application of cell culture is exceptionally essential in life science and medical science. Cell culture technique is an excellent tool its applications are [1, 39, 40] as a). Production of pharmaceutical biochemicals b). Embryological study (CEF cell culture) c). Recombinant biomaterials (rDNA) and vaccine manufacturing, testing of the drug, drug sensitivity, and cytotoxicity of cell d). Production of human and animal vaccines (primary chicken fibroblast cells e). Manufacturing of immunotherapy f). Production of different enzymes, hormones (synthetic), immunobiological (like monoclonal antibodies, interleukins, lymphokines), and anticancer agents g). Cell culture is an excellent way to teach cell biology study h). Production of agricultural products like milk, (cultured) meat, fragrances i). In the microbiological study (Virus propagation, virology) j). Genetic engineers and biotechnologist are utilizing it within their field of research k). The aging, toxic compound study, cell morphology, cell physiology, and the study of mutagenesis cell culture have great impact.

7. CELL HANDLING, STORAGE AND CHALLENGES

There are a lot of measures that should be taken to handle cells smoothly. Smooth and gentle handling of the cell gives better outcomes. An SOP must be defined clearly in the handling procedure of a cell. Clean area, BSL, pressure control of the cell culture room, personal hygiene, general laboratory management, gowning, skilled manpower can play a vital role in this regard. Generally, cell incubate at 36–37°C in a 5% CO2 sometimes CO2 level 4–10%, and time required based on cell types, research methodology. For the short preservation (few hours <2 hours) of the cell need normal freezing and for about day-long preservation required -20°C with growth media or cryoprotectant agent. Cryopreservation is required for the yearlong storage of cells. The temperature logbook is an essential document in the lab. Any fluctuation, problems in power supply, user entry of fridge/ cold room, and incubator are inevitably helpful to handle and maintain cells. Besides lock, access control or password system in laboratory, incubator, and fridge might be helpful in this regard. Resulting in easily identify the problems if any. Vigorous pipetting during cell harvesting, splitting cause cell damage sometimes causes cell death. A few toxic and harmful substances are eluted from the microfilters during sterilization. That's why practicing the sterile technique strictly and selecting cultural instruments carefully. It is recommended that washing all the instruments with the culture medium immediately before performing cell culture. Nevertheless, other types of contamination happened from plastic instruments or trace elements, even in water that affects the cells in culture [11]. Viruses, bacteria, mycoplasma, and endotoxins contamination may appear due to rough handling and haphazard performance. Strict environmental control is necessary to check cross-contamination in cell culture [41].

8. QUALITY CONTROL, GMP AND CGMP

Quality control (QC) involves ensuring the quality and validity of research or laboratory materials such as cells, medium, PBS, Polymerase chain reaction (PCR), or RT-PCR reports. The Quality Control department and personnel are crucial. Every life science researcher should be familiar with the current Good Manufacturing Practice (cGMP), quality control procedure, and system. QC, GMP, and cGMP share similar aims and operations. GMP and cGMP clearly outline rules, regulations, and guidelines for all areas of the product's lifespan and manufacturing process, including product creation, raw materials selection, production process, testing, storage, and distribution. Global regulatory bodies such as the FDA, MHRA, WHO, and EUDRALEX-UK have endorsed all cGMP guidelines. Recently, GMP has been designated as "cGMP." The "c" represents "current," emphasizing the need for methodologies, systems, and procedures to be regularly updated to comply with the latest regulations. [42]. cGMP guidelines ensure excellent quality at each step of work. It acts as a safeguard for the quality, purity, strength and identity of cells and related products. This guideline helps to construct a consolidated quality control and management system, helps to hold raw materials quality, helps to develop good SOPs, helps to a build a system of product quality investigation and deviation measurement, and take part in the building of reference laboratories for testing. The necessity of cGMPs are always keep in mind because it is very much important to establish QC. Numerous pharmaceutical companies are executing modern and comprehensive quality control systems and risk management system as a result they hold their quality above the average standards [43]. Quality control (QC) is employed in the GMP framework. Using a broad array of analytical techniques, the QA (Quality assurance) team will identify and quantify all the factors deemed critical for any specific product. Biological products require extensive analysis for characterization. There are some cations related to QC [44] such as-

- identity (confirms correct material);
- quantity (confirms dose parameter);
- purity (confirms material has correct purity);
- impurity (confirms product safety);
- potency (confirms activity);
- sterility including adventitious agents (confirms material is sterile).

9. OVERALL CHALLENGES: CELL CULTURE

Culturing techniques for CEF cells or BHK-21 or others is always challenging. The followings are the most important points

by maintaining these points researchers or laboratory personnel easily can overcome any possible challenges. The points that need to pay attention such as a). Lab requirement is in the first, set up all the required materials, chemicals, instruments before the work or research b). Novice personnel should be avoided c). Skilled manpower and staff should be appointed for the work and as a trainer for new one d). Interpretation of work daily e). Establish a strong QC department f). Store all the true data and metadata because it is important to identify problems and also help to design a research g). Calibration of pH meter regular basis is mandatory and pH level follow up because cell health depends on it h). Need to take necessary precautions during culture media preparation i). Maintain and check the required CO2 percentage based on cell j). Incubation time and temperature vary from cell to cell and it may be vary based on research work k). Transportation process, medium, time, precautions should be describing in transportation SOP l). Pay attention during thawing of the cell because it causes stress on the cell resulting cell might be death m). More careful and be cautious during preservation of cells, changing of media, passaging, trypsinization, harvesting of cells, mycoplasma contamination, and antibiotics treatment (for bacterial contamination) because these are the fundamental steps in cell culture technology. A non-technical or unskilled or novice personnel can easily ruin all the effective work at a time during the above points. That's why need to be more cautious about this n). Certificate of Analysis (COA) is essential for all the work like media preparation, pH level check, cell morphology, cell count, splitting, transportation, cell purchase or sell, and so on. It reduces the risk O). Develop a standard operating procedure (SOP) based on worldwide reference laboratory guidelines. Creating a SOP for preparing effective SOPs and work guidelines can help reduce issues and contribute to addressing concerns. Adhering to Current Good Manufacturing Practice (cGMP), maintaining thorough documentation of lab work, providing training for researchers and staff, validating work, and frequently calibrating instruments can effectively address issues and manage challenges efficiently.

10. EFFECTIVENESS OF MAINTAINING LOGBOOK, TRAINING AND SOPS

Understanding the distinction between a laboratory logbook and a laboratory notebook is crucial for efficient laboratory operations. A logbook records specific information in a structured format regarding the management and use of equipment, users, time of usage, purpose of use, and relevant observations. The research facility notebook or diary comprises comprehensive experimental details, readings, calculation results, and extra data such as graphs, spectra, or chromatograms. All activities in a research facility involving testing equipment, laboratory environmental conditions, weighing balance logs, material consumption, chemicals, supplier details, and supplies are documented as proof records. Vary the types of logbooks based on the nature of the work. Both notebooks and logs are essential for decision-making, problem identification, and rectifying inaccurate results. During an audit, a logbook helps to convince the auditor because it provides evidence-based data. Regular maintenance, conducted by skilled personnel, calibration, and servicing of equipment prolong the instrument's lifespan, and these records are kept as evidence for safe operations. SOP-based work has an equivalent effect on research. On the other hand, training is significant for the advancement of the skills and knowledge of laboratory staff. In a training log the topic of training, date and time, name of trainer and trainees, effectiveness, and also the performance of individuals is reserved [34, 45]. Does every detail in the SOP provide clear instructions on how to operate an instrument, conduct a test, or use a machine? Who will be eligible to perform? How can one effectively monitor and maintain records for various activities such as laboratory work and tests? When and how should calibration be performed? Who provides certification and through what process? How to utilize chemicals? SOP is essential. It is advisable to establish precise protocols for managing and regulating laboratory supplies, as outlined in detailed Standard Operating protocols (SOP).

11. COCISE SUMMARY

The significance and utilization of cell culture in recent years are difficult to summarize briefly. Cell culture technology is a widely established approach in the study of molecular biology and life science research. Cell culture procedures are in significant demand in biopharmaceutical efforts due to their high practicality. This document outlines the theoretical framework and fundamental ideas related to cell culture, specifically focusing on animal cells, BHK-21 cells, and CEF cell culture. The preceding conversation provides a comprehensive understanding of the cell culture technology, its applications, and aids in developing problem-solving skills and confidence in this field. Ultimately, all the steps involved in cell culture provide a basic understanding of the process.

12. AT A GLANCE

Cell culture is a method performed in vitro where cells, tissues, or organs of animal origin are artificially cultivated in a controlled environment that includes a culture medium, specific CO2 levels, pH indicators, and temperature regulation to ensure the cells' survival and proper growth. Cell culture is extensively utilized for studying cell growth, normal functioning, identifying growth factors, developing viral vaccines, utilizing recombinant DNA (rDNA) technology, and conducting immunobiological research, distinguishing it as the most commonly used method among organ culture and primary explant culture. Cell culture procedures are in high demand in the pharmaceutical business due to their high practicality. Animal cell culture is utilized in laboratory research to investigate several aspects such as cytotoxicity of novel medications, metabolic studies, aging, therapeutic proteins, effects of pharmaceuticals and harmful chemicals on cells, and mutagenesis and carcinogenesis. Cell culture faces numerous challenges, with Mycoplasma infection being a significant concern. One antibiotic is usually ineffective at eradicating mycoplasma during cell culture. Additionally, factors such as culture media, pH indicator, incubation, cryopreservation, thawing, passaging of cells, and trypsinization significantly influence cell culture. This chapter will enhance the reader's comprehension of cell culture and its applications, advancing their research in virology and cell culture. This chapter discusses cell count, cell suspension, CCF measurement, MOI (Multiplicity of Infection), and cell infection. The reader will eventually have a thorough understanding of cell culture.

ABBREVIATIONS

ATCC	American Type Culture Collection
BHK	Baby Hamster Kidney
BME	Basal Medium
BOD	Biological Oxygen Demand
BSC	Biosafety Cabinet
BSL	Biosafety Level
CCF	Cell Culture Flask
$CCID_{50}$	Cell Culture Infectious Dose 50
CDC	Center for Disease Control and Prevention
CEF	Chicken Embryo Fibroblast
cGMP	Current Good Manufacturing Practice
CHO	Chinese Hamster Ovary
COA	Certificate of Analysis
CPAs	Cryoprotective Agents
DAPI	4⍰,6-diamidino-2-phenylindole
DMEM	Dulbecco's Modified Eagle's Medium
DMSO	Dimethyl Sulphoxide
DT	Doubling Time (Population)
EC_{50}	Effective Concentration
EDTA	Ethylenediaminetetraacetic Acid
EID_{50}	Embryo Infective Dosage

ELD_{50}	Embryo Lethal Dosage
ELISA	Enzyme-Linked Immunosorbent Assay
EMEM	Eagle's Minimum Essential Medium
FBS	Fetal Bovine Serum
FDA	Food and Drug Administration
FFA	Focus forming Assay
FMD	Foot and Mouth Disease
GCCP	Good Cell Culture Practice
GM	Growth Media
GMP	Good Manufacturing Practice
HA	Hemagglutination Assay
Hpi	hours post-inoculation
IC_{50}	Concentration of an Inhibitor
IMDM	Iscove's Modified Dulbecco's Medium
IPA	Isopropyl Alcohol
ISO	International Organization for Standardization
LD_{50}	Lethal Dose 50
MHRA	Medicines & Healthcare Regulatory Agency
MOI	Multiplicity of Infection
NIH	National Institute of Health, USA
OECD TG	Organization for Economic Co-Operation and Development Test Guide Line
PBS	Phosphate Buffer Saline
PCR	Polymerase Chain Reaction
PDL	Population Doubling Level
PFU	Plaque-Forming Unit
PPE	Personal Protective Equipment
QA	Quality Assurance
QC	Quality Control
R&D	Research and Development
rDNA	Recombinant Deoxyribonucleic Acid
RDS	Room Data Sheet

ROS	Reactive Oxygen Species
Rpm	Rotation Per Minute
RT-PCR	Reverse Transcription Polymerase Chain Reaction
SOP	Standard Operating Procedures
SPF	Specific Pathogen Free
$TCID_{50}$	Tissue Culture Infectious Dose 50
TEM	Transmission Electron Microscopy
UPS	Uninterrupted Power Supply
WHO	World Health Organization

REFERENCES

1. Farzaneh M, Attari F, Mozdziak PE, Khoshnam SE. The evolution of chicken stem cell culture methods. British Poultry Science, 2017, 58 (6): 681-686. DOI:10.1080/00071668.2017.1365354
2. Mandrioli M, Monti M, Tedeschi R. A practical guide to insect cell cultures: establishment and maintenance of primary cell cultures. HALTERES, 2015, 6: 132-141
3. Freshney RI. Culture of Animal Cells: A Manual of Basic Technique and Specialized Applications. 2010, Wiley, John & Sons, Inc, New Jersey
4. DGDA (Directorate General of Drug Administration). Guidelines on Animal Vaccine. 2018, Ministry of Health and Family Welfare Government of the People's Republic of Bangladesh
5. Kantardjieff A, Zhou W. Mammalian Cell Cultures for Biologics Manufacturing. In: Mammalian Cell Cultures for Biologics Manufacturing. 2014, Springer, Heidelberg, p. 1
6. Auniņš JG. Viral Vaccine Production in Cell Culture. In: Encyclopedia of Industrial Biotechnology: Bioprocess, Bioseparation, and Cell Technology. Wiley, New York 2010
7. Stoker M, Macpherson I. Syrian hamster fibroblast cell line Bhk21 and its derivatives. Nature, 1964, 203:1355-1357

8. Butler M. In: Animal Cell Culture and Techonology, Internl, T.J (ed.). Padstow, Cornwall, 2nd edition, UK, 2003, 10-27

9. Zhang, XZ, Chen Y, Huang HL, Xu DL, Ren CB, Liu BT, Su S, Tang ZX. Pseudorabies virus induces viability changes and oxidative stress in swine testis cell-line. Pakistan veterinary journal, 2013, 33: 438-441

10. Griffiths JB. Mammalian cell culture reactors, scale-up. In: Flickinger MC (ed) Encyclopedia of industrial biotechnology: bioprocess, bioseparation, and cell technology. Wiley, New York, 2010, 1-13

11. Yao T, Asayama Y. Animal-cell Culture media: History, characteristics, and current issues. Reprod Medical Biology. 2017; 16: 99-117. https://doi.org/10.1002/rmb2.12024

12. Brunner D, Frank J, Appl H, Schoffl H, Pfaller W, Gstraunthaler G. Serum-free cell culture: the serum-free media interactive online database. ALTEX, 2010, 27: 53-62

13. Bhatia S, Naved T, Sardana S. Introduction to Pharmaceutical Biotechnology, Volume 3 Animal tissue culture and biopharmaceuticals. Chapter 1. Introduction to animal tissue culture science. IOP Publishing Ltd, 2019. doi:10.1088/2053-2563/aafac0ch1

14. Verma A, Verma M, Singh A. Animal tissue culture principles and applications. Animal biotechnology: tools and techniques. Chapter 14. Elsevier Inc.2020, 269-293.https://doi.org/10.1016/B978-0-12-811710-1.00012-4

15. Nema R, Khare S. An animal cell culture: Advance technology for modern research. Advances in Bioscience and Biotechnology. 2012; 3: 219-226

16. Pohjanpelto P. Putrescine transport is greatly increased in human fibroblasts initiated to proliferate. The Journal of cell biology, 1976; 68: 512-520

17. Ibtisham F, Zhao Y, Nawab A, Liguang H, Wu J, Xiao M, Zhao Z, An L. The Effect of High Temperature on Viability, Proliferation, Apoptosis and Anti-oxidant Status of Chicken Embryonic Fibroblast Cells. Brazilian Journal of Poultry Science, 2018, 20 (3): 463-470

18. Dubey AK, Lavanya L, Sadananda D, Gouthami K, Elfansu K, Singh A. Inferences of Carbon Dioxide in Present-Day Cell Culture Systems: An Unacknowledged Problem and Perspectives. Austin Therapeutics, 2021, 6(1): 1033

19. Maria TL, Eric KJ, Ehrhardt RO. Cryopreservation and Thawing of Mammalian Cells. In: eLS. John Wiley & Sons, Ltd: Chichester, 2014. DOI: 10.1002/9780470015902.a0002561.pub2

20. Choi J, Bischof JC. Cooling rate dependent biophysical and viability response shift with attachment state in human dermal fibroblast cells. Cryobiology, 2011, 63: 285-291

21. Hunt CJ, Armitage SE, Pegg DE. Cryopreservation of umbilical cord blood: 2. Tolerance of CD34+ cells to multimolar dimethyl sulphoxide and the effect of cooling rate on the recovery after freezing and thawing. Cryobiology, 2003, 46: 76-87

22. Naaldijk Y, Staude M, Fedorova V, Stolzing A. Effect of different freezing rates during cryopreservation of rat mesenchymal stem cells using combinations of hydroxyethyl starch and dimethylsulfoxide. BMC Biotechnology, 2012, 12: 49

23. Ragai RM, Robin DH. Human Cell Culture Protocols, Methods in Molecular Biology, vol. 806, DOI 10.1007/978-1-61779-367-7_1

24. ATCC. ANIMAL CELL Animal Cell Culture Guide. www.atcc.org

25. Wayne M, Maria YL. Rolf TO. Ehrhardt. Cryopreservation and Thawing of Cells. Current Protocols in Immunology. Wiley Online Library. John Wiley & Sons, Inc.2012. DOI: 10.1002/0471142735.ima03gs99

26. Mather JP, Roberts PE. Introduction to Cell and Tissue Culture: Theory and Technique. 1998

27. Bhatia S, Naved T, Sardana S. Introduction to Pharmaceutical Biotechnology, Volume 3 Animal tissue culture and biopharmaceuticals. Chapter 1. Introduction to animal tissue culture science. IOP Publishing Ltd, 2019. doi:10.1088/2053-2563/aafac0ch1

28. Science Gateway. Cell Biology Protocols. https://www.sciencegateway.org/protocols/cellbio/cell/moipfu.htm

29. Kärber G. Beitrag zur kollektiven Behandlung pharmakologischer Reihenversuche. Archiv f. Experiment. Pathol. U. Pharmakol. Springer-Verlag, 1931, 162 (4): 480-483. doi:10.1007/BF01863914

30. Kaufmann SH, Kabelitz D. Methods in Microbiology: Immunology of Infection. Academic Press. 2002, vol-32. ISBN 0-12-521532-0

31. Liang T, Liu X, Cui S, Qu S, Wang D, Liu N, Wang F, Ning K, Zhang B, Zhang D. Generation of a reliable full-length cDNA of infectious Tembusu virus using a PCR-based protocol. Virus Res. 2016, 213, 255-259

32. Shahla Shahsavandi, Mohammad Majid Ebrahimi, Ashraf Mohammadi, Nima Zarrin Lebas. Impact of chicken-origin cells on adaptation of a low pathogenic influenza virus. Cytotechnology, 2013, 65:419-424. DOI 10.1007/s10616-012-9495-5

33. Coecke S. Guidance on good cell culture practice: A Report of the Second ECVAM Task Force on good cell culture practice. ATLA Alternatives to Laboratory Animals, 2005,33 (3): 261-287

34. OECD. Test No. 432: In Vitro 3T3 NRU Phototoxicity Test, OECD Guidelines for the Testing of Chemicals, Section 4, OECD Publishing, Paris, 2004, http://dx.doi.org/10.1787/9789264071162-en

35. Maneesha S, Inamdar, Lyn Healy, Sinha A, Stacey G. Global Solutions to the Challenges of Setting up and Managing a Stem

Cell Laboratory. Stem Cell Rev and Rep, 2012, 8:830-843. DOI 10.1007/s12015-011-9326-7

36. Coecke S, Balls M, Bowe G, Davis J, Gstraunthaler G, Hartung T. Guidance on good cell culture practice (GCCP). A report of the second ECVAM task force on good cell culture practice. ATLA, 2005, 33: 1-27

37. Drexler HG, Upho CC. Mycoplasma Contamination of Cell Cultures: Incidence, Sources, Effects, Detection, Elimination, Prevention. Cytotechnology, 2002, 39: 75

38. RegMedNet. Biological Industries. 2017. https://www.regmednet.com/10-tips-for-preventing-mycoplasma-contamination/

39. Maguire G. Therapeutics from Adult Stem Cells and the Hype Curve. ACS Medicinal Chemistry Letters, 2016, 7 (5): 441 3. doi:10.1021/acsmedchemlett.6b00125

40. Prieto D, Aparicio G, Sotelo-Silveira JR. Cell migration analysis: A low-costlaboratory experiment for cell and developmental biology courses using keratocytes from fish scales. Biochemistry and Molecular Biology Education, 2017, 45 (6):475-482. doi: 10.1002/bmb.21071

41. Ragai R, Mitry, Robin D, Hughes. Human Cell Culture Protocols, Methods in Molecular Biology, vol. 806, DOI 10.1007/978-1-61779-367-7_1

42. BI (Biological Industries). cGMP: GeneralCell Culture, Stem Cell Research. An Interview with Ravid Grimberg. 2019. https://blog.bioind.com/cgmp-what-every-life-science-researcher-should-know-about-current-good-manufacturing-practice/

43. FDA. Facts About the Current Good Manufacturing Practices (CGMPs) https://www.fda.gov/drugs/pharmaceutical-quality-resources/facts-about-current-good-manufacturing-practices-cgmps

44. Green A, Sharpe G. Good Manufacturing Practice for Cell Culture Processing. Medicines from Animal Cell Culture Edited by G. Stacey and J. Davis. John Wiley & Sons, Ltd, 2007. ISBN: 978-0-470-85094-7

45. OECD. OECD Principles on Good Laboratory Practice, OECD Series on Principles of Good Laboratory Practice and Compliance Monitoring, No. 1, OECD Publishing, Paris 1998. http://dx.doi.org/10.1787/9789264078536-en

CHAPTER 2

PRIMARY CELL CULTURE MODELS IN PRECLINICAL RESEARCH OF PROSTATE AND BLADDER CANCER

1. INTRODUCTION: AN OVERVIEW

Cancer is a leading cause of death globally, responsible for roughly 10 million deaths in 2020[1]. Prostate cancer and bladder cancer incidence rates are rising globally. Prostate cancer ranks as the second most common cancer and the fifth leading cause of death among men, as reported by GLOBOCON 2018. Bladder cancer is prevalent in men, ranking sixth in frequency and ninth in terms of cancer-related deaths [2]. Various treatments such as radiation, chemotherapy, and hormonal therapy are available, but they are linked to negative side effects and a reduced quality of life after treatment. Therefore, there is a necessity to design efficient, secure, and economically feasible anticancer medications. Prostate cancer and bladder cancer are complex diseases influenced by several molecular, environmental, and hereditary variables, making it challenging to comprehend the progression mechanism[3]. Cancer research has advanced significantly in recent years, however numerous challenges persist[4]. Only 7% of possible anticancer treatments get approved, a significantly lower rate compared to medications for other disorders[5]. To enhance this percentage, it is crucial to validate medications in preclinical investigations and allow them to progress to phase I clinical trials[6].

Experimental models play a crucial role in cancer research. The model must be replicable and accurately represent the illness stage under investigation, as well as simulate the behavior of the disease in patients [4]. Cell lines are in vitro model systems, a necessary tool, in not only the search for new substances showing antitumor activity but additionally for assessing their effectiveness. They are widely used in different fields of medical research and pharmaceutical companies. Presently pharmaceutical industries mostly rely on in vitro models like two dimensional (2D), three dimensional (3D), boyden's chamber (to study chemotaxis and assessment of cell motility) [7], micro fluidic systems (It is small devices that can provide a specific fluid flow, constant temperature, fresh medium, flow pressure and chemical gradients which is same as in vivo systems to study migration and invasion [8], 3D bioprinting (mimics the processes that occurs in tumor micro environment) [9, 10]. Main reason for accepting in vitro model is it's physiological relevance, it helps in improving the understanding of prognosis and treatment, it provides accuracy and it is also a low cost screening tool for researchers [11]. The usefulness of in vitro models is primarily linked to their ability to provide an indefinite source of biological material for experimental purposes. The in vivo model involves animals which provide valuable information to understand many aspects in development of disease and initial development of drugs such as toxicity, corrosion and drug activity [12]. But from past two decades, alternatives have been sought due to the fact that animals do not effectively model humans in in vivo conditions, as it shows unexpected responses like anatomical variation and also difficulty in extracting quantitative mechanistic data in the studies. Mathematical models are also used in the cancer research to analyze tumor growth and progression, and helps in predicting the effects of some therapies [13]. Different clinical setting, cancer resistance and switching to another treatment, existence of unknown biological details these issues can affect the mathematical models [14, 15, 16]. Computer simulation is a model used in cancer research to assess complicated multi-scale cancer progression and therapeutic effects. However, it may be limited in less frequent malignancies due to a lack of data, resulting in challenges in validation and accuracy [17].

Each model used in cancer research has advantages and disadvantages. Therefore, factors such as cost, time, experimental design, and data analysis should be taken into account when choosing a model for producing an anticancer medicine. It is essential to select more efficient preclinical platforms for evaluating the anticancer drugs [18]. In-vitro tumor models provide first screening for possible anticancer medications and help avoid ineffective treatments from progressing to preclinical animal research[19]. This chapter examines the primary characteristics of cancer cell cultures, outlines various methods for their selection and handling, and summarizes the diverse studies that can enhance the comprehension of preclinical treatment processes for prostate and bladder cancer.

2. PRIMARY CELL CULTURE

Primary cell culture is considered the gold standard testing platform for in-vitro oncology research since it more closely reflects the tumor state compared to routinely used cell lines [20]. Scientists frequently employ this potent technique to examine biological characteristics and functions of individual cells in a regulated setting [21]. Cell culture studies have positively impacted the early stages of medication development. Unlike animal research, the in vitro method necessitates a low medication dose and a brief reaction period, which are distinctive features of in vitro cell culture techniques [22]. Primary cell culture is also known as ex-vivo. Primary cells are derived directly from tissue and cultivated under optimal conditions, making them closely resemble the in vivo state and display typical physiological characteristics. It sustains communication between cancerous and normal elements [23]. Primary cell culture is considered an ideal model system for conducting investigations on metabolism, aging, signaling, and the impact of medicines and hazardous chemicals on cells. Primary cells are non-immortalized and non-transformable as they imitate the appearance of living model and hence these can help to model 3D tissues. In this culture, cells will grow in 3D aggregates and presents interesting application [24].

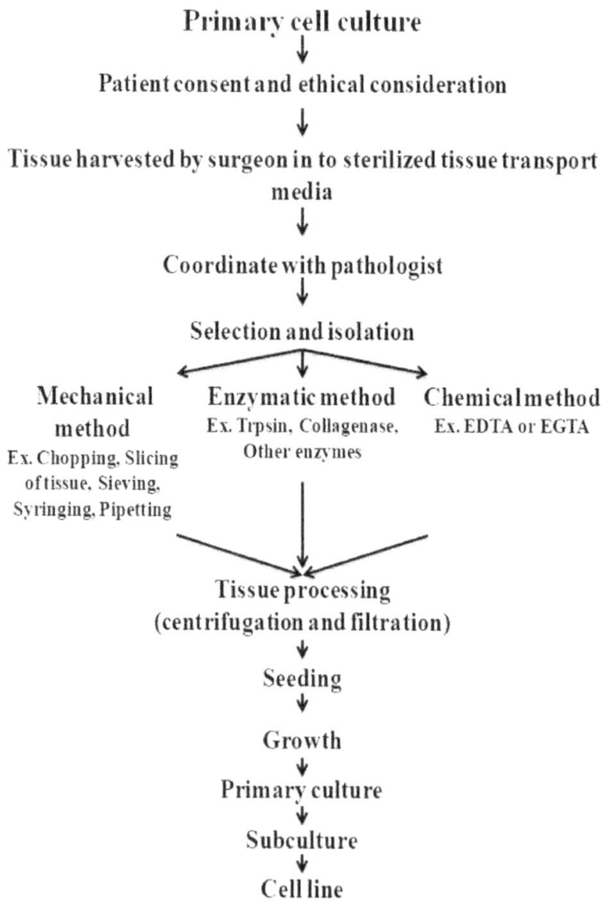

Figure 1. Flowchart showing the process of primary cell culture technique.

It helps for detection, isolation, growth and developmental stages of viruses and helps to analyze the mode of infection. Drug candidate and its toxicity screening rely on results of early-stage in vitro cell based assays. Particularly in pharmaceutical industry primary cell culture is used to synthesize verity of biomolecules in high scale, various research project on cell-based product are developed. It is alternative for animal model to test effect of drugs and cosmetics [25]. There are minimal technological challenges associated with primary cell cultivation. Culturing could be challenging if the surgical material's quality is low. Early beginning of cell senescence poses challenges in maintaining an adequate number of cell passages, however researchers have made numerous efforts to address this issue[26, 27].

Primary cell culture is separated into adherent cell culture and suspension cell culture. Adherent cell culture involves cells organized in a single layer that stick to the surface of the culture flask. Adherent cells typically originate from organ tissues. Growth is constrained by surface area and must be removed from the surface prior to cell sub-culturing. Viral vaccine, gene therapy and cell therapies depend only on adherent cell culture. Suspension cells originate from peripheral circulation and can grow without needing to connect to a surface. They do not adhere to the surface area of the culture flask. The cells are suspended and their proliferation is constrained by the cell concentration. Figure 1 illustrates the steps of primary cell cultivation.

3. ISOLATION OF CELLS

Prior to proceeding with any more tissue treatments, it is crucial to conduct all tissue processing within a biosafety cabinet and adhere to correct sterilizing methods [28]. Now moving towards cell isolation, it is a process where one or more specific cells are isolated from heterogenous cell mixture. Isolation of primary cells from cancer cells is an important phenomenon of cell culture biology as they are more reliable sources to understand the human cell. There are many standard protocols available for culturing the normal and neoplastic cells [28]. Human prostate and bladder are composed of many cell types which can be isolated and cultured. Hierarchy of the epithelium has been reviewed most [29, 30]. There are 3 main epithelial lineages namely neuroendocrine, basal and luminal. Prostatic homeostasis is mainly depends upon the epithelial cells and stromal cells; stromal cells guides to the epithelium cell for their dedifferentiation, proliferation and also progression of carcinogenesis [31].

How can one differentiate between cancerous and non-cancerous cells when cells lack identifying tags? Distinct cell markers (antigens) can differentiate between malignant and non-cancerous cells. ARA70, also known as Androgen receptor-associated protein 70, is a cell marker that shows high expression in normal primary cultures but lower levels in prostate cancer cell lines. [32]. Many cell markers are available depending upon cell type which is listed in Table 1. Also in bladder cancer depending upon cell type there are different CD (Cluster of Differentiation) cell markers which are depicted in Table 2 [38]. Identification of stem cell marker has uncovered a cellular hierarchy of epithelium during development and in response to injury [39]. Cell markers (Antigens) have specific antibodies and these have to be evaluated histochemically. These reactions are evaluated by specific kits which are available in market (Table 3). For isolation, first tissues will be collected from prostate cancer patients and bladder cancer patients who are undergoing biopsy. This collection of tissues needs to be well coordinated between urology, pathology and the investigator and it has to proceed for primary culture within 2 hours after collection of tissues [33, 34, 35, 36, 37, 39, 40, 41].

Following collection, tissue should be stored in a sterile container containing HBSS (Hanks' balanced salt solution) with HEPES (Hydroxyethyl piperazineethanesulfonic acid) at 4°C for 2 hours to enhance cell viability. Hematoxylin and Eosin (H&E) stain is utilized in the histopathology lab to evaluate tumor cells in the dissected material. Three approaches are available for isolating a single cell suspension from tissue dissociation produced after sugary. Chemical, mechanical, and enzymatic methods are presented in Table 4.

Table 1. Biomarkers for Prostate cell culture.

Cell marker	Cell type
Cytokeratin 5 [33]	Basal cell
CD59 [34]	Basal cell
c-Met [35]	Basal cell
CD95 [34]	Basal cell
Cytokeratin 8 [36]	luminal cell
CD9 [37]	luminal cell
15-LOX-2 [36]	luminal cell
CD24 [34]	luminal cell

Table 2. Biomarkers for bladder cell culture.

Cell marker	Cell type
CD9 [38]	Urothelial
CD104+ [38]	Basal
CD13+ [38]	Stromal cells of the lamina propria
CK5 [37]	Basal
p63 [37]	Basal
CK8 [37]	Luminal

Table 3. Primary antibodies for prostate and bladder epithelial cultures [38, 40].

Antigen	Antibodies	Cell type
Cytokeratin 8/18 (Prostate)	Mouse IgG1	Luminal cells and intermediate cells
High molecular weight cytokeratin (Prostate)	Mouse IgG1	Intermediate cells
Trop2-APC (Prostate)	Mouse IgG2a	Epithelial cells
CD49f-PE(Prostate)	Rat IgG2a	Epithelial cells
CD + 9 (Bladder)	antibody clone M-L13	Urothelial
CD104 (Bladder)	clone 439-9B	Basal

Table 4. Mechanisms for isolation of cells.

Method	Mechanism	Advantages	Disadvantages
Chemical Method	By using EDTA or EGTA it binds with cations and disrupt the intracellular bonds [42]	Easy and cost saving	It does not adequately dissociate all types of tissue
Mechanical Method	Cutting, scratching, the tissues in to small pieces in order to separate the cells and wash it with gentle agitation [42]	It is a rapid technique works best for loosely associated tissue. Correct temperature should be maintained for enzyme.	Decreases in the surviving capacity of the cell, incision of scissor, scalpel for cutting, scratching can damage the cell.
Enzymatic Method	Enzymes to cut or digest tissue pieces in free cells. Combination of enzyme also can be used Ex.-Collagenase, Trypsin, Hyluronidase [43].	It has great specificity with specific enzymes	Enzyme dissociation can modify proteins on cell surface

While this is the initial stage of primary culture, there is yet no established protocol for it. There are numerous options accessible. and process tissue, it must be minced using an autoclaved scalpel or scissor. If the tissue weighs between 1 and 20 grams, a semi-automated dissociator can be utilized. Manual procedure must be carried out in ice-cold Phosphate-buffered saline (PBS). The commercially available formulation exhibited a 10% higher viability compared to collagenase I, II, and IV. Table 5 [33]. In another study mechanical and enzymatic method has been used. In mechanical method, they used lacerate and scalpels and in enzymatic method collagenase type I and hyaluronidase type I enzymes with medium agitation at 37°C for 18 hours was used [34]. EDTA (Ethylenediaminetetraacetic acid)/Trypsin mixture used with 5 minutes of incubation in 37°C degree for prostate tissue [35]. Both the mechanisms, mechanical disaggregation with disposable disaggregator and enzymatic by collagenase and trypsin used for prostate tissue [36, 37]. In some cases, trypsin/EDTA 1:5 solution and incubation for 15 minutes for bladder tissue was used. In some studies for dissociation of bladder tissue 1:1 collagenase II and dispase enzymes are used at 37°C for 12 hours [40, 41, 42, 43]. It is necessary to monitor the tissue digestion process every 2 hours by gently stirring the digestion mixture and assessing the cell viability under a microscope[50, 51, 52, 53].

Table 5. Basic components of media and their functions.

Sl. No.	Components	Function
1	Buffering systems Natural buffer system Chemical buffers system	CO_2 balances the pH,5 to 10% Co_2 incubation, non-toxic and cost effective [44] HEPES, buffering capacity7.2–7.4, in high concentration it is toxic, costly [45, 46]
2	Inorganic salt	Maintains osmotic balance and membrane potential by providing sodium, potassium, and calcium ions [47]
3	Amino acids	Require for proliferation of cells and provides nitrogen Ex. L-glutamine provides NAD,NADPH and serve as secondary source of energy in metabolism [48].
4	Carbohydrates	Carbohydrates are major source of energy, most of the media use glucose and Galactose.
5	Proteins and peptides	Major proteins and peptides in media are albumin, fibronectin and tranferrin. Albumin helps to remove toxic substances from cell culture media
6	Fatty acids and lipids	Particularly added when serum free media is used [44].
7	Vitamins	Important for cell growth and proliferation as cells cannot produce sufficient amount, need to provide through culture media [44].
8	Trace elements	Copper, Zink, Selenium required in trace amount for proper growth and many biological process
9	Antibiotics	Controls growth of bacteria and fungi ex.Penicillin, Streptomycin, fluconazole [49].
10	Serum in media	Serum is a complex mixture provides all above elements. Ex. Fetal and Calf bovine serum [44].

4. TISSUE PROCESSING

After digestion, cells are filtered using a strainer to separate detritus. Next, cell micro-clumps are washed two or three times with PBS or HBSS and then centrifuged [54]. The cell pellet obtained from centrifugation is resuspended in 2 ml of culture medium. Quantify the live cells using a hemocytometer or the tryptophan dye exclusion method [55]. Cell viability can also be assessed by measuring intracellular adenosine triphosphate levels using a commercially available kit [56]. Immunohistochemistry and immunofluorescence techniques are employed to pinpoint, recognize, and measure cells based on specific cell surface markers [57].

5. CULTURE MEDIA

Cell pellets obtained after centrifugation should be transferred to a micro well plate or flask containing culture medium. It creates an artificial environment for cell growth. The essential components of culture media are regulated temperature, a surface for cell attachment, a nutrient-rich environment, and an incubator for pH maintenance [45]. The primary stage in culturing is selecting the culture media. The composition often includes amino acids, vitamins, inorganic salts, glucose, hormones, growth factors, and attachment factors that supply energy and support cell cycle completion. Commercial cell media for primary epithelial cancer cells are less effective than tissue-specific primary cell media generated in a laboratory (Table 6) [46, 47, 48, 49].

Table 6. Components of media from different studies.

SL No.	Cell type	Components of media
1.	Fibroblast cell culture [28]	DMEM media with 7.2 pH + Fetal calf serum (FCS) + 100 U/ml penicillin, 100 µg/ml streptomycin +1% amphotericin B added in culture media. culture plate incubated at 37°C a humidified chamber of 95% air and 5% CO_2
2.	Prostate cancer (Bone metastatic variant) [29]	DMEM Glutamax +4.5 g/L D-Glucose with pyruvate +10%FCII +1% penicillin–streptomycin+37°C with 5% CO_2.

SL No.	Cell type	Components of media
3.	Bladder cancer (Epithelial cells) [58]	EMEM (ATCC) +10% FCS +1% penicillin + streptomycin + humidified incubator at 37°C with 5% CO_2.
4.	Prostate cancer (Epithelial cells) [59]	KSFM medium+ + 25 mM HEPES +1% penicillin + streptomycin +0.5 mg/mL fungizone +100 mg/mL gentamicin +37°C, 5% CO 2 humidified incubation.
5.	Epithelial cells [40]	serum-free RPMI 1640 without phenol red+ penicillin 100 IU/mL+ streptomycin 100 μg/mL+ metronidazole 1 μg/mL+ amphotericin B 2.5 μg/mL + gentamicin 20 μg/mL + 37°C and 5% CO_2 for 6 days.
6.	Bladder cancer (Urothelial cells) [50]	Glutamine + insulin + Phosphoethanolamine+ ethanolamine + hydrocortisone+ transferrin + EGF + BPHE+5% FBS +5% CO_2 at 37°C
7.	Prostate cancer (Epithelial cells) [60]	DMEM + Glucose +100 U/ml penicillin +100 mg/ml streptomycin sulfate+0.29 mg/ml glutamine + Euroclone
8.	Prostate cancer (Epithelial cells) [61]	DEME /Ham's (1:1) + BSA (0.01%) + FAS (2%) + Epidermal growth factor (10 ng/mL) + (insulin-transferring-selenium-1%) + hydrocortison (0.5 μg/mL) + Tryiodotyronone(1 nM) + phosphpethenolamine(0.1 mM) + choleratoxin (50 ng/mL) + fibronectin (100 ng/mL) + futine (20 μg/mL) + penicillin/streptomycin (100 U/mL,100 μg/mL) + R1881(0.1 nM).

The selection of culture media is crucial for obtaining meaningful results in experiments. The choice of media is determined by the cell type, intended use, and available resources [62]. High-quality culture medium is essential for obtaining relevant research data from primary cultures. Alternatively, commercially available media can be used to overcome limitations in primary cell culture, such as cell quantity (Table 7) [58, 59, 60, 61].

Table 7. Commercially available media for epithelial cells [44, 63].

Sl.No.	Name of media	Supplier
1.	Human Endothelial-SFM	Life Technologies
2.	Endothelial Basal Media	Sigma Aldrich
3.	EndoGRO-LS Complete Media Kit	MilliporeSigma
4.	HUVEC Basal Medium CB HUVEC	AllCells

5.	Endothelial Cell Medium	ScienCell
6.	Epithelial cell medium	ScienCell
7.	EpiGRO primary epithelial cells	MilliporeSigma
8.	RPMI 1640	Sigma Aldrich
9.	αMEM	ThermoFischer
10.	k- SFM	ThermoFischer

Cell survival post-isolation is crucial and relies heavily on skilled manipulation and optimal culture conditions. Cultural requirements vary based on the specific cell type. Cell growth must be monitored for 11 to 12 days. Extra media, Fetal Bovine Serum (FBS), and antibiotics must be supplied to prevent contamination. The culture media needs to be replaced every 2 to 3 days. [60]. Initially apoptosis is 5% from 0 to 1 day but as days will pass apoptosis rate will increases from 7 to 14 days. But functional validity of benign and prostate cancer cells was 5 days after confirming it with histochemically, biochemical and by immunohistochemical assay [63]. Use of serum free culture media with low calcium condition increases the longevity of the cell. Cryopreservation (Preservation of structurally intact cells) can be achieved by adding 10% FBS (Fetal bovine serum) and 10% DMSO (Dimethyl sulfoxide) in 80% confluence primary cell culture [64]. Each day morphological changes have to benoted. Normal cells get counted every day and cancer cells get counted every 2 days [65]. Cell viability is determined by trypsin blue dye, equal volume of PBS and trypsin blue dye allowed to sit on cells for few minutes then to count the cells samples are loaded on hemocytometer, cells scored as leaving or dead based on uptake of tryptophan blue dye [66]. Once confluence reaches to 80 to 90% it has to get counted by Neubauer camera at 1:2 dilution with tryptophan dye exclusion, MTT assay (3-(4,5-dimethylthiazol-2-yl)-2,5-diphenyl tetrazolium bromide) can be used to determine cell viability [67]. A cell growth curve can be generated from the graph to determine the point at which cell viability rises in the culture. After cells are cultivated appropriately to meet the investigator's requirements, they can be passaged and characterized.

6. CANCER CELL LINES

The predominant prostate cancer cell lines used in research investigations include PC-3, DU145, and LNCaP, which were established in the 1970s and 1980s[68, 69, 70]. T24 cells, a cell line derived from transitional cell carcinoma of the bladder, are commonly used in bladder cancer research [71]. Immortalized cancer cell lines may not accurately predict real cancer behavior in preclinical research due to their adaptation to 2D monolayer growth conditions [72, 73].

7. CELL CULTURE MODELS

Animal models have traditionally been used for studying various forms of cancer throughout the last thirty years [74]. Animal studies have limitations such as the absence of efficient high-throughput drug screening, prolonged testing duration, and ethical debates surrounding animal experimentation. Cell culture is a common alternative to animal investigations, with techniques falling into two main categories: 2D and 3D methods [75]. The utility of primary cancer cell models in preclinical investigations for cancer research and medication discovery has increased in recent years. Primary cell cultures offer a valuable model for studying both normal and malignant biological processes. Recent studies have focused on examining behaviors associated with carcinogenesis, including apoptosis, proliferation, adhesion, differentiation, migration, senescence, invasion, angiogenesis, and several metabolic pathways. The primary culture's heterogeneity of cell populations replicates the tumor microenvironment, interactions between malignant and healthy cells, and crosstalk, which cannot be achieved using cell lines[76].

Research indicates that cellular reactions to drug treatments in 3D cell culture are substantial and closely resemble in vivo architecture, as opposed to 2D cell culture. An advanced and successful method that has shown improvement in cell-based screening during the initial stages of drug discovery is the assessment of cancer cell viability using 3D culture. This assay is very valuable for evaluating the cytotoxic impact of substances that could induce cell death. It is crucial for determining the number of viable cells at the conclusion of each experiment. Cell viability assay is subsequently followed by cell proliferation, cell migration, and cell signaling assays [77, 78, 79].

There are currently various chemically distinct anticancer medicines accessible. Testing chemicals must impede the development and proliferation of cancer cell lines to be considered as a possible anticancer agent. This will impede the signaling pathways by introducing or removing a candidate gene, therefore halting the advancement of the tumor to lethal phases. Antiproliferative studies were conducted on the DU-145 prostate cancer cell line both in vitro and in vivo utilizing Salvia miltiorrhiza Bunge [80]. An additional illustration to support this concept is a study conducted on the PC-3 cell line. Cancer cells typically exhibit elevated levels of Transferrin Receptors (TfR) to facilitate greater absorption of iron compared to normal cells. Cancer cells require an increased level of intracellular free iron for their development and proliferation. Anhydrodihydroartemisinin (ADHA) was utilized to suppress PC-3 cell lines via a caspase-dependent route[81].

8. APPLICATIONS: CELL CULTURE IN DRUG DISCOVERY AND SCREENING

8.1 Drug candidate identification

The target identification and validation step is frequently the bottleneck in preclinical drug research. 3D cell cultures can help identify the molecular changes that drive cancer development and speed up the process of identifying and confirming targets. This is because the gene expression patterns in 3D models more closely resemble those in living organisms compared to 2D monolayer models[81]. A study found that CXCR7 and CXCR4 were co-expressed in LNCaP, DU145, and PC3 cell lines in 2D culture. PC3 cells cultivated in 3D using Matrigel showed increased expression of both receptors, indicating that blocking CXCR7/CXCR4 could help regulate prostate tumor development and progression[82].

8.2 Toxicity profiling

Cultured cancer cells are effective for evaluating drug-induced toxicity and identifying appropriate medications and techniques for specifically targeting various forms of cancer. Studying the impact of medication responses on metabolic signaling pathways or candidate genes is beneficial due to advancements in drug screening procedures over the past decade. A study examined characteristics including vascularization and perfusion of antineoplastic medicines on human T24 bladder cancer. (83]. It allowed in the understanding of basic paracrine signaling mechanisms that regulates tissue homeostasis, development of new methods for urinary bladder reconstruction and tissue engineering, and generation of models of malignant and benign diseases. This study suggested that the use of 3D urinary bladder cultures could be a possible approach in clinical practice to select for the best antineoplastic drug for each patient and to investigate the effect of drug combinations or new antineoplastic drugs [84, 85]. The below (Figure 2) suggests how assay-guided treatment can be useful in choosing the best active drug for an individual patient.

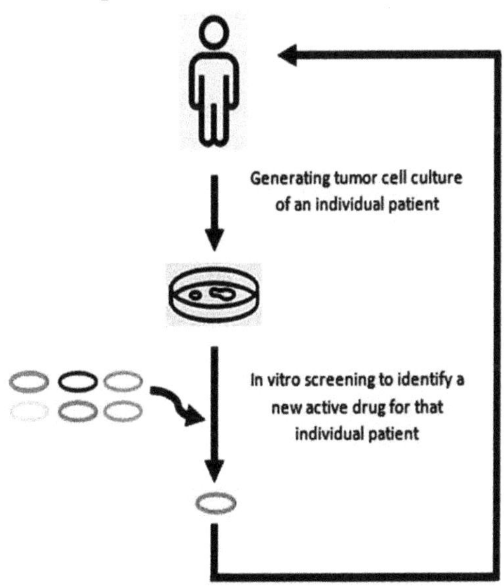

Figure 2. Assay-guided treatment choosing the best active drug for an individual patient.

8.3 Testing anticancer activities

The most valuable in vitro model for studying anticancer activity is cell culture. Various treatments such as radiation, chemotherapy, hormone therapy, and innovative experimental medicines can be assessed. Plant extracts can be used to investigate anti-cancer properties such anti-inflammatory effects and destabilization of membranes that facilitate invasion and migration. Leaf extracts of Leea indica were utilized to investigate antioxidant and anticancer properties in vitro on DU-145 and PC-3 human prostate cancer cell lines. An instance of drug combination is the synergistic impact of cisplatin and sunitnib malate-based chemotherapy on T24, 5637, and HT1376 human urinary bladder cancer cell lines [87].

9. CONCISE SUMMARY

Primary cell cultures are utilized in toxicology, virology, drug screening, genetic engineering, gene therapy, genetic counseling, and cancer research, with a primary focus on serving as a model system. It offers an excellent model system for investigating fundamental cell biology and biochemistry, the impact of disease-causing agents on cells, the effects of medications on cells, and the processes that initiate aging and apoptosis. Primary cell culture is an ideal model for first preclinical studies aimed at comprehending cancer inside an in-vitro setting. Primary cell culture is considered the gold standard for cell line investigations due to its ability to offer a wider range of cell types from a larger pool of patients, without introducing artificial genetic mutations, while still maintaining consistent phenotypes throughout the culture. It engages both clinicians and researchers in studying cultural influences to better comprehend therapeutic limitations and methodological shortcomings in laboratories, enabling strategies to address these challenges. Various models exist in cancer research, each with distinct capabilities and limitations. Primary cell culture complexity can be attributed to suboptimal tissue quality, improper collection methods, and inadequate cultured conditions, which can lead to reduced cell viability. Managing primary cell culture is challenging. Proper collection, selection of isolation method, and culture media based on tissue type can boost cell viability with the guidance of a pathologist. Given the current clinical approach to precision medicine, patient-derived cancer models represent a potent example in cancer research. The 3D model system is currently an emerging technology. Primary cell culture can be utilized to simulate 3D culture. In the future, 3D culture may potentially replace the in vivo model system in technological advancements. This chapter summarizes many aspects of primary cell culture, including tissue selection, isolation procedures, culturing media, and cell maintenance post-culture. It provides a summary of several studies aimed at enhancing the comprehension of preclinical treatment procedures for prostate and bladder cancer.

10. AT A GLANCE

The global rise in diagnosed cases of prostate and bladder cancer highlights the crucial need for the creation of efficient, secure, and cost-effective antitumor medications. Obtaining clinical approval based on preclinical investigations is crucial for medications to progress to phase I clinical trials. Cell lines are commonly utilized in several areas of medical research, particularly in basic cancer research and drug discovery, as in vitro model systems. They are principally valuable for their capacity to offer an unlimited supply of biological material for research. Authenticated cancer cell lines, when subjected to suitable circumstances and controls, preserve the majority of the genetic characteristics of the original cancer. Studies conducted in the early stages of drug development focused on toxicity, corrosion, and drug activity using animals. However, in the last twenty years, alternatives have been explored because animals do not accurately represent human in vivo conditions and unexpected reactions occur in these studies. Over 100 million animals were utilized and billions of dollars were expended on animal toxicity research. Cell culture studies have significantly contributed to the early stages of drug discovery and are highly valued for their ability to offer convenient and direct access to tissues for evaluation. In contrast to animal experiments, in vitro cell culture procedures are characterized by lower costs, the requirement for less medication, and a shorter reaction time. In vitro tumor models are essential for identifying new compounds with anticancer properties and evaluating their efficacy. This chapter examines the primary characteristics of cancer cell cultures, outlines various methods for their selection and handling, and summarizes the diverse studies that can enhance the comprehension of preclinical treatment processes for prostate and bladder cancer.

ABBREVIATIONS

ADHA	Anhydrodihydroartemisinin
BPHE	Brazed Plate Heat Exchangers
BSA	Bovine Serum Albumin
CD cell markers	Cluster of Differentiation cell markers
DMEM	Dulbecco's Modified Eagle Medium
DMSO	Dimethyl sulfoxide
DU145	Duke University 145
EDTA	Ethylenediaminetetraacetic acid
EGTA	Ethylene glycol tetraacetic acid
EMEM	Eagle's Minimum Essential Medium
FBS	Fetal Bovine Serum
FCS	Fetal Calf Serum
FGF	Fibroblast Growth Factors
HBBS	Hanks' balanced salt solution
HEPES	Hydroxyethylpiperazineethanesulfonic acid
IgMI	Imunoglobuline M Imunostain
KSFM	Keratinocyte Serum Free Medium
LNCaP	Lymph Node Carcinoma of the Prostate
PBS	Phosphate buffer saline
RPMI 1640	Roswell Park Memorial Institute (RPMI) media
TfR	Transferrin Receptors

REFERENCES

1. Cancer. https://www.who.int/news-room/fact-sheets/detail/cancer

2. Ferlay J, Colombet M, Soerjomataram I, Mathers C, Parkin DM, Piñeros M, Znaor A, Bray F. Estimating the global cancer incidence and mortality in 2018: GLOBOCAN sources and methods. International journal of cancer. 2019; 144(8):1941-1953

3. Rhim JS. In vitro human cell culture models for the study of prostate cancer. Prostate Cancer and Prostatic Diseases. 2000; 3(4):229-235

4. Nascimento-Gonçalves E, Ferreira R, Oliveira PA, Colaço BJ. An overview of current alternative models for use in the context of prostate cancer research. Alternatives to Laboratory Animals. 2020; 48(2):58-69

5. Solovyeva VV, Kitaeva KV, Rutland CS, Rizvanov AA. Cell Culture Based in vitro Test Systems for Anticancer Drug Screening. Frontiers in Bioengineering and Biotechnology.;8

6. Loucera C, Esteban-Medina M, Rian K, Falco MM, Dopazo J, Peña-Chilet M. Drug repurposing for COVID-19 using machine learning and mechanistic models of signal transduction circuits related to SARS-CoV-2 infection. Signal transduction and targeted therapy. 2020; 5(1):1-3

7. Falasca M, Raimondi C, Maffucci T. Boyden chamber. InCell Migration 2011 (pp. 87-95). Humana Press

8. Ruzycka M, Cimpan MR, Rios-Mondragon I, Grudzinski IP. Microfluidics for studying metastatic patterns of lung cancer. Journal of nanobiotechnology. 2019; 17(1):1-30

9. Kang HW, Lee SJ, Ko IK, Kengla C, Yoo JJ, Atala A. A 3D bioprinting system to produce human-scale tissue constructs with structural integrity. Nature biotechnology. 2016; 34(3):312-319

10. Truong D, Fiorelli R, Barrientos ES, Melendez EL, Sanai N, Mehta S, Nikkhah M. A three-dimensional (3D) organotypic microfluidic model for glioma stem cells–Vascular interactions. Biomaterials. 2019; 198:63-77

11. Katt ME, Placone AL, Wong AD, Xu ZS, Searson PC. In vitro tumor models: advantages, disadvantages, variables, and selecting the right platform. Frontiers in bioengineering and biotechnology. 2016; 4:12

12. C Gonçalves V, JLL Pinheiro D, de la Rosa T, G de Almeida AC, A Scorza F, A Scorza C. Propolis as a potential disease-modifying strategy in Parkinson's disease: cardioprotective and neuroprotective effects in the 6-OHDA rat model. Nutrients. 2020; 12(6):1551

13 .Zazoua A, Wang W. Analysis of mathematical model of prostate cancer with androgen deprivation therapy. Communications in Nonlinear Science and Numerical Simulation. 2019; 66:41-60

14. Blackard CE, Byar DP, Jordan WP. Vacurg (1973) Orchiectomy for advanced prostatic carcinoma. Urology.;1(6):553

15. Hirata Y, Bruchovsky N, Aihara K. Development of a mathematical model that predicts the outcome of hormone therapy for prostate cancer. Journal of theoretical biology. 2010; 264(2):517-527

16. Rich JN. Cancer stem cells in radiation resistance. Cancer research. 2007; 67(19):8980-8984

17. Pai RK, Van Booven DJ, Parmar M, Lokeshwar SD, Shah K, Ramasamy R, Arora H. A review of current advancements and limitations of artificial intelligence in genitourinary cancers. American Journal of Clinical and Experimental Urology. 2020;8(5):152

18. Imamura Y, Mukohara T, Shimono Y, Funakoshi Y, Chayahara N, Toyoda M, Kiyota N, Takao S, Kono S, Nakatsura T, Minami H. Comparison of 2D-and 3D-culture models as drug-testing

platforms in breast cancer. Oncology reports. 2015; 33(4):1837-1843

19. Stevens JL, Baker TK. The future of drug safety testing: expanding the view and narrowing the focus. Drug discovery today. 2009; 14(3-4):162-167

20. Janik K, Popeda M, Peciak J, Rosiak K, Smolarz M, Treda C, Rieske P, Stoczynska-Fidelus E, Ksiazkiewicz M. Efficient and simple approach to in vitro culture of primary epithelial cancer cells. Bioscience reports. 2016;36(6): e00423

21. Ager-Wick E, Hodne K, Fontaine R, von Krogh K, Haug TM, Weltzien FA. Preparation of a high-quality primary cell culture from fish pituitaries. JoVE (Journal of Visualized Experiments). 2018 (138): e58159

22. ŞAHİN ŞH, MESUT B, ÖZSOY Y. Applications of Cell Culture Studies in Pharmaceutical Technology. ACTA Pharmaceutica Sciencia.;55(3)

23. Yin L, Qin G, Qian HZ, Zhu Y, Hu W, Zhang L, Chen K, Wang Y, Liu S, Zhou F, Xing H. Continued spread of HIV among injecting drug users in southern Sichuan Province, China. Harm reduction journal. 2007; 4(1):1-7

24. Joseph JS, Malindisa ST, Ntwasa M. Two-dimensional (2D) and three-dimensional (3D) cell culturing in drug discovery. Cell Culture. 2018; 2:1-22

25. https://bioscience.lonza.com/lonza_bs/IN/en/what-are-primary-cells-and-how-do-they-differ-from-cell-lines [Internet] 2021

26. Stoczynska-Fidelus E, Piaskowski S, Bienkowski M, Banaszczyk M, Hulas-Bigoszewska K, Winiecka-Klimek M, Radomiak-Zaluska A, Och W, Borowiec M, Zieba J, Treda C. The failure in the stabilization of glioblastoma-derived cell lines: spontaneous in vitro senescence as the main culprit. PloS one. 2014; 9(1): e87136

27. Zieba J, Ksiazkiewcz M, Janik K, Banaszczyk M, Peciak J, Piaskowski S, Lipinski M, Olczak M, Stoczynska-Fidelus E, Rieske P. Sensitivity of neoplastic cells to senescence unveiled under standard cell culture conditions. Anticancer research. 2015; 35(5):2759-2768

28. Ranganathan K, Kavitha L. Oral epithelial dysplasia: Classifications and clinical relevance in risk assessment of oral potentially malignant disorders. Journal of oral and maxillofacial pathology: JOMFP. 2019; 23(1):19

29. Shen MM, Abate-Shen C. Molecular genetics of prostate cancer: new prospects for old challenges. Genes & development. 2010; 24(18):1967-2000

30. Wang Y, Hayward SW, Cao M, Thayer KA, Cunha GR. Cell differentiation lineage in the prostate. Differentiation. 2001 Oct 1;68(4-5):270-279

31. Taylor RA, Risbridger GP. Prostatic tumor stroma: a key player in cancer progression. Current cancer drug targets. 2008; 8(6):490-497

32. Tekur S, Lau KM, Long J, Burnstein K, Ho SM. Expression of RFG/ELE1α/ARA70 in normal and malignant prostatic epithelial cell cultures and lines: regulation by methylation and sex steroids. Molecular Carcinogenesis: Published in cooperation with the University of Texas MD Anderson Cancer Center. 2001; 30(1):1-3

33. Brawer MK, Peehl DM, Stamey TA, Bostwick DG. Keratin immune reactivity in the benign and neoplastic human prostate. Cancer research. 1985; 45(8):3663-3667

34. Liu AY, Peehl DM. Characterization of cultured human prostatic epithelial cells by cluster designation antigen expression. Cell and tissue research. 2001 Sep;305(3):389-397

35. You X, Yu HM, Cohen-Gould L, Cao B, Symons M, Woude GF, Knudsen BS. Regulation of migration of primary prostate

epithelial cells by secreted factors from prostate stromal cells. Experimental cell research. 2003; 288(2):246-256

36. Van Leenders G, Dijkman H, Hulsbergen-van de Kaa C, Ruiter D, Schalken J. Demonstration of intermediate cells during human prostate epithelial differentiation in situ and in vitro using triple-staining confocal scanning microscopy. Laboratory investigation. 2000; 80(8):1251-1258

37. Sfakianos JP, Daza J, Hu Y, Anastos H, Bryant G, Bareja R, Badani KK, Galsky MD, Elemento O, Faltas BM, Mulholland DJ. Epithelial plasticity can generate multi-lineage phenotypes in human and murine bladder cancers. Nature communications. 2020;11(1):1-6

38. Liu AY, Vêncio RZ, Page LS, Ho ME, Loprieno MA, True LD. Bladder expression of CD cell surface antigens and cell-type-specific transcriptomes. Cell and tissue research. 2012; 348(3):589-600

39. Strand DW, Aaron L, Henry G, Franco OE, Simon W. populations. 2017;91(224):139-51. DOI10.1016/j.diff.2015.10.013.Isolation

40. Failli A, Consolini R, Legitimo A, Spisni R, Castagna M, Romanini A, Crimaldi G, Miccoli P. The challenge of culturing human colorectal tumor cells: establishment of a cell culture model by the comparison of different methodological approaches. Tumori Journal. 2009; 95(3):343-347

41. Caspar A, Mostertz J, Leymann M, Ziegler P, Evert K, Evert M, Zimmermann U, Brandenburg LO, Burchardt M, Stope MB. In vitro cultivation of primary prostate cancer cells alters the molecular biomarker pattern. in vivo. 2016; 30(5):573-579

42. Castell JV, Gómez-Lechón MJ. Liver cell culture techniques. Hepatocyte transplantation. 2009:35-46

43. Chen S, Principessa L, Isaacs JT. Human prostate cancer initiating cells isolated directly from localized cancer do not form prostaspheres in primary culture. The Prostate. 2012; 72(13):1478-1489

44. Meenakshi A. Cell culture media: a review. Mater Methods. 2013; 3:175

45. Shipman Jr C. Evaluation of 4-(2-hydroxyethyl)-1-piperazineëthanesulfonic acid (HEPES) as a tissue culture buffer. Proceedings of the Society for experimental biology and medicine. 1969; 130(1):305-310

46. Zigler JS, Lepe-Zuniga JL, Vistica B, Gery I. Analysis of the cytotoxic effects of light-exposed HEPES-containing culture medium. In Vitro Cellular & Developmental Biology. 1985; 21(5):282-287

47. Činátl J. Inorganic-organic multimolecular complexes of salt solutions, culture media and biological fluids and their possible significance for the origin of life. Journal of theoretical biology. 1969; 23(1):1-0

48. Pasieka AE, Morgan JF. Glutamine metabolism of normal and malignant cells cultivated in synthetic media. Nature. 1959; 183(4669):1201-1202

49. Perlman D. Use of antibiotics in cell culture media. InMethods in enzymology 1979 Jan 1 (Vol. 58, pp. 110-116). Academic Press

50. Roberson KM, Yancey DR, Padilla-Nash H, Edwards DW, Nash W, Jacobs S, Padilla GM, Larchian WA, Robertson CN. Isolation and characterization of a novel human bladder cancer cell line: BK10. In Vitro Cellular & Developmental Biology-Animal. 1998; 34(7):537-544

51. Goldstein AS, Drake JM, Burnes DL, Finley DS, Zhang H, Reiter RE, Huang J, Witte ON. Purification and direct transformation of epithelial progenitor cells from primary human prostate. Nature protocols. 2011; 6(5):656-667

52. Grainger A. Difficulties in tracking the long-term global trend in tropical forest area. Proceedings of the National Academy of Sciences. 2008; 105(2):818-823

53. Navone NM, Olive M, Troncoso P. Isolation and culture of prostate cancer cell lines. InCancer Cell Culture 2004 (pp. 121-132). Humana Press

54. Kim FJ, Campagna A, Khandrika L, Koul S, Byun SS, Bokhoven AV, Moore EE, Koul H. Individualized medicine for renal cell carcinoma: establishment of primary cell line culture from surgical specimens. Journal of endourology. 2008; 22(10):2361-2366

55. Ali MY, Anand SV, Tangella K, Ramkumar D, Saif TA. Isolation of primary human colon tumor cells from surgical tissues and culturing them directly on soft elastic substrates for traction cytometry. Journal of visualized experiments: JoVE. 2015(100)

56. Sharma S, Neale MH, Di Nicolantonio F, Knight LA, Whitehouse PA, Mercer SJ, Higgins BR, Lamont A, Osborne R, Hindley AC, Kurbacher CM. Outcome of ATP-based tumor chemosensitivity assay directed chemotherapy in heavily pre-treated recurrent ovarian carcinoma. BMC cancer. 2003; 3(1):1-0

57. Maity B, Sheff D, Fisher RA. Immunostaining: detection of signaling protein location in tissues, cells and subcellular compartments. Methods in cell biology. 2013; 113:81-105

58. Van de Merbel AF, van der Horst G, van der Mark MH, van Uhm JI, van Gennep EJ, Kloen P, Beimers L, Pelger R, van der Pluijm G. An ex vivo tissue culture model for the assessment of individualized drug responses in prostate and bladder cancer. Frontiers in oncology. 2018; 8:400

59. Sato T, Clevers H. Primary mouse small intestinal epithelial cell cultures. InEpithelial cell culture protocols 2012 (pp. 319-328). Humana Press, Totowa, NJ

60. Papini S, Rosellini A, De Matteis A, Campani D, Selli C, Caporali A, Bettuzzi S, Revoltella RP. Establishment of an organotypic in vitro culture system and its relevance to the characterization of human prostate epithelial cancer cells and their stromal

interactions. Pathology-Research and Practice. 2007; 203(4):209-216

61. Zhang W, van Weerden WM, de Ridder CM, Erkens-Schulze S, Schönfeld E, Meijer TG, Kanaar R, van Gent DC, Nonnekens J. Ex vivo treatment of prostate tumor tissue recapitulates in vivo therapy response. The Prostate. 2019; 79(4):390-402
62. Clifford WJ, Anellis A, Ross Jr EW. Evaluation of media, time and temperature of incubation, and method of enumeration of several strains of Clostridium perfringens spores. Applied microbiology. 1974; 27(4):784-792
63. Wang R, Chu GC, Wang X, Wu JB, Hu P, Multani AS, Pathak S, Zhau HE, Chung LW. Establishment and characterization of a prostate cancer cell line from a prostatectomy specimen for the study of cellular interaction. International journal of cancer. 2019;145(8):2249-2259
64. Maund SL, Nolley R, Peehl DM. Optimization and comprehensive characterization of a faithful tissue culture model of the benign and malignant human prostate. Laboratory Investigation. 2014; 94(2):208-221
65. Speirs V, Green AR, Walton DS, Kerin MJ, Fox JN, Carleton PJ, Desai SB, Atkin SL. Short-term primary culture of epithelial cells derived from human breast tumours. British journal of cancer. 1998; 78(11):1421-1429
66. Samsel L, Zaidel G, Drumgoole HM, Jelovac D, Drachenberg C, Rhee JG, Brodie AM, Bielawska A, Smyth MJ. The ceramide analog, B13, induces apoptosis in prostate cancer cell lines and inhibits tumor growth in prostate cancer xenografts. The Prostate. 2004; 58(4):382-393
67. Raxworthy MJ. Animal cell culture: A practical approach: Edited by RI Freshney. pp 248. IRL Press, Oxford. 1986.£ 22 or£ 14 (pbk) ISBN 0-947946-62-4 or 0-947946-33-0 (pbk)

68. Horoszewicz JS. The LNCaP cell line-a new model for studies on human prostatic carcinoma. Prog Clin Biol Res. 1980; 37:115-132

69. Mickey DD, Stone KR, Wunderli H, Mickey GH, Vollmer RT, Paulson DF. Heterotransplantation of a human prostatic adenocarcinoma cell line in nude mice. Cancer research. 1977; 37(11):4049-4058

70. Kaighn ME, Narayan KS, Ohnuki Y, Lechner JF, Jones LW. Establishment and characterization of a human prostatic carcinoma cell line (PC-3). Investigative urology. 1979;17(1):16-23

71. Bubenik J, Barešová M, Viklický V, Jakoubkova J, Sainerova H, Donner J. Established cell line of urinary bladder carcinoma (T24) containing tumour-specific antigen. International journal of cancer. 1973; 11(3):765-773

72. Hamacher R, Bauer S. Preclinical models for translational sarcoma research. Current opinion in oncology. 2017; 29(4):275-285

73. Gillet JP, Varma S, Gottesman MM. The clinical relevance of cancer cell lines. Journal of the National Cancer Institute. 2013; 105(7):452-458

74. Ravi M, Paramesh V, Kaviya SR, Anuradha E, Solomon FP. 3D cell culture systems: advantages and applications. Journal of cellular physiology. 2015; 230(1):16-26

75. Joseph JS, Malindisa ST, Ntwasa M. Two-dimensional (2D) and three-dimensional (3D) cell culturing in drug discovery. Cell Culture. 2018; 2:1-22

76. Miserocchi G, Mercatali L, Liverani C, De Vita A, Spadazzi C, Pieri F, Bongiovanni A, Recine F, Amadori D, Ibrahim T. Management and potentialities of primary cancer cultures in preclinical and translational studies. Journal of translational medicine. 2017;15(1):1-6

77. Comley J. 3D cell culture, easier said than done. Drug Discov World. 2010:25-41

78. Riss TL, Moravec RA, Niles AL, Duellman S, Benink HA, Worzella TJ, Minor L. Cell viability assays. Assay Guidance Manual [Internet]. Eli Lilly & Company and the National Center for Advancing Translational Sciences. 2016

79. Bae WJ, Choi JB, Kim KS, Ha US, Hong SH, Lee JY, Hwang TK, Hwang SY, Wang ZP, Kim SW. Inhibition of Proliferation of Prostate Cancer Cell Line DU-145 in vitro and in vivo Using Salvia miltiorrhiza Bunge. Chinese journal of integrative medicine. 2020; 26(7):533-538

80. Ahmad F, Sarder A, Gour R, Karna SK, Arora P, Kartha KR, Pokharel YR. Inhibition of prostate cancer cell line (PC-3) by anhydrodihydroartemisinin (ADHA) through caspase-dependent pathway. EXCLI journal. 2020; 19:613

81. Ghosh S, Spagnoli GC, Martin I, Ploegert S, Demougin P, Heberer M, Reschner A. Three-dimensional culture of melanoma cells profoundly affects gene expression profile: A high density oligonucleotide array study. Journal of cellular physiology. 2005; 204(2):522-531

82. Lovitt CJ, Shelper TB, Avery VM. Advanced cell culture techniques for cancer drug discovery. Biology. 2014; 3(2):345-367

83. Zips D, Thames HD, Baumann M. New anticancer agents: in vitro and in vivo evaluation. in vivo. 2005; 19(1):1-7

84. Varley CL, Southgate J. Organotypic and 3D reconstructed cultures of the human bladder and urinary tract. In3D Cell Culture 2011 (pp. 197-211). Humana Press

85. Johannsen M, Gneveckow U, Thiesen B, Taymoorian K, Cho CH, Waldöfner N, Scholz R, Jordan A, Loening SA, Wust P. Thermotherapy of prostate cancer using magnetic nanoparticles: feasibility, imaging, and three-dimensional

temperature distribution. European urology. 2007; 52(6):1653-1662

86. Ghagane SC, Puranik SI, Kumbar VM, Nerli RB, Jalalpure SS, Hiremath MB, Neelagund S, Aladakatti R. In vitro antioxidant and anticancer activity of Leea indica leaf extracts on human prostate cancer cell lines. Integrative medicine research. 2017; 6(1):79-87

87. Arantes-Rodrigues R, Pinto-Leite R, Fidalgo-Gonçalves L, Palmeira C, Santos L, Colaço A, Oliveira P. Synergistic effect between cisplatin and sunitinib malate on human urinary bladder-cancer cell lines. BioMed research international. 2013; 2013

CHAPTER 3

APPLICATIONS OF TWO-DIMENSIONAL AND THREE-DIMENSIONAL CELL CULTURE

1. INTRODUCTION: AN OVERVIEW

Cell culture is the growing of cells in a controlled artificial environment separated from their natural habitat[1]. This technology is commonly utilized for studying cell and molecular biology, drug screening, toxicity analysis, investigating the impact of specific genes on diseases, and conducting cancer research. Because of their distinctive characteristics, they have also been optimized for screening and developing medicinal molecules including vaccines and recombinant proteins. Cell culture offers the benefit of producing consistent and replicable data[2].

Drug discovery is a protracted and laborious process that involves multiple rounds of testing and refinement. This includes identifying the target, discovering leads, validating in pre-clinical stages, and conducting clinical trials [3]. Hence, it is crucial to gather data on the biological activity, biochemical processes, toxicity, and off-target interactions of pharmacological compounds during the initial phases of drug discovery. Cell-based assays conducted in vitro are ineffective in comprehending the impact of medications on cells during the initial stages of drug discovery, which enhances the likelihood of developing drugs with high efficacy and safety[4].

Two-dimensional (2D) cell culture was established many decades ago and has since become the predominant method of cell culture in various industries. This conventional method has been widely employed for drug screening because of its cost-effectiveness and ease of usage. Nevertheless, the challenge of replicating the in vivo environment limits its application [5]. 2D cell cultures grow in a single layer in controlled flat environments like glass or polystyrene flasks. This layer consists of live proliferating cells due to the removal of dead cells from the surface. This results in a consistent distribution of nutrients and growth factors in the media, allowing cells to access them and proliferate more rapidly than they would naturally. The cells' shape is altered, causing them to seem flattened and elongated relative to their natural in vivo state [6]. Additionally, the interactions between cells and between cells and the extracellular environment are altered when compared to normal tissue in the tumor[7].

There has been a recent increase in interest in three-dimensional (3D) cell culture in biomedical research and drug development due to its high-throughput accuracy and advanced in vitro models [8]. They have been widely utilized to study cell morphology, cell communication, and the cellular microenvironment that closely resembles the in vivo conditions. 3D cultures develop as clusters or aggregates known as spheroids, either with or without a matrix [9]. There is a variation in nutrition levels within these spheroids, causing cells on the surface to multiply more than those in the center. Due to varying proliferation rates, cells within spheroids are typically found in multiple stages of the cell cycle, including proliferating, quiescent, hypoxic stages and necrotic cells. [10].

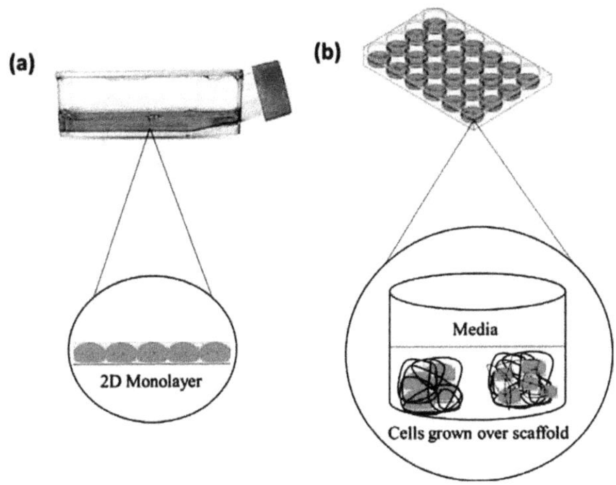

Figure 1. Simplified sketch of 2D cell culture (a) and 3D cell culture (b).

2. IN VITRO APPLICATIONS OF 2D CULTURE

2.1 In vitro cytotoxicity assays and tissue-engineered tissue models

Cytotoxicity assays are commonly used for in vitro toxicology and pharmacology studies for the screening the effect of chemicals and drugs on the cultured cells. There are different assays available for measuring cytotoxicity namely- the colony-forming assay and dye inclusion or exclusion such as neutral red and trypan blue assay is the most significantly used. Cytotoxicity assays can be broadly divided into the following categories- (i) Assays based on metabolism, (ii) Adenosine triphosphate (ATP) Bioluminescence Assay and (iii) Assays based on the release of enzymes.

Assays based on metabolism generally include the 3-(4,5-dimethythiazol2-yl)-2,5-diphenyl tetrazolium bromide (MTT) assay and its alternatives such as 2,3-Bis-(2-Methoxy-4-Nitro-5-Sulfophenyl)-2H-Tetrazolium-5-Carboxanilide (XTT),3-(4,5-dimethylthiazol-2-yl)-5-(3-carboxymethoxyphenyl)-2-(4-sulfophenyl)-2H-tetrazolium(MTS) and sulforhodamine B (SRB) assay. Due to rapid, quantitative, versatile, and highly reproducibility of MTT, it is widely used in large-scale, anti-tumor drug-screening program. MTT is a quantitative colorimetric assay that quantifies the reduction of yellow tetrazolium dye by mitochondrial succinate dehydrogenase to purple insoluble formazan crystals by the NADPH dependent cellular oxidoreductase enzymes [12]. The crystals are dissolved in an appropriate solvent. The absorbance is then recorded using a spectrophotometer to analyze the cell viability wherein the crystals get accumulated in the viable cells due to their impermeability to the cell membrane.

ATP bioluminescence An assay is utilized to quantify the ATP levels, which are tightly controlled in metabolically active living eukaryotic cells contrasted to dead cells, where ATP levels decrease owing to ATPase activity. This assay involves a luciferase enzyme that uses ATP energy to convert luciferin into oxyluciferin, resulting in light. Hence, luminescence could be utilized for quantifying ATP levels. Enzyme-based assays are more significant as they quantify the products produced by deceased cells[13].

An enzyme-based assay involves measuring LDH leakage by monitoring the conversion of lactate to pyruvate in the presence of LDH, which results in the reduction of NAD to NADH and a change in absorbance at 340 nm. [14]. Research in cancer and cell biology is greatly dependent on in vitro assays and models. This help in understanding the various responses of the cultured cells when exposed to different conditions. Tissue-engineered in vitro tissue models serve as an alternative to in vivo animal studies to study the physiology of various diseases. Example of tissue-engineered in vitro models includes Skeletal Muscle Models [15], blood–brain barrier model [16], aneurysm models [17] and the Pre-vascularized Human Vaginal Mucosa model [18].

2.2 Cell migration assay

Cell migration is well known for its significant role in embryonic morphogenesis, cancer invasion and metastasis, immune responses, tissue formation, and angiogenesis [19]. Cell migration primarily occurs in two forms: single-cell migration and collective cell migration. Individual cell movement is controlled by cytoskeletal function independently of interactions with adjacent cells. This movement is crucial for embryonic development, immunological response, and early metastases. In collective cell migration, a group of cells maintains their connections and polarity as they move together. The wound healing test, often known as the scratch assay, is a 2D in vitro method utilized to investigate collective cell movement. This experiment involves creating a scratch on the dense cell monolayer, leading to the development of a gap or wound. The progress of cell migration is tracked by capturing images of the cells moving at set time intervals. The images are utilized to assess the rate of wound healing and quantify cell movement. Time-lapse microscopy enables live-cell imaging for a comprehensive analysis of cell movement patterns [20]. To minimize the impact of cell proliferation on gap filling, data are collected over a 24-hour period, which may be adjusted based on the specific cell line being studied.

The transwell test, also known as the Boyden Chamber experiment, examines how individual cells respond to different chemo-attractants. This assay is suitable for both adherent and non-adherent cells. Cells are placed in a serum-free medium on one side of a porous membrane and their migration through the pores to the other side is examined. Cell migration can be measured by tallying the cells that have moved through the membrane towards the area with a greater concentration of chemoattractant [21]. One limitation of this experiment is the challenge of observing the cells and their shape as they go through pores because of the transient nature of cells [22].

2.3 In vitro tumorigenicity assay

Cancer is a significant health burden in both developing and wealthy countries, causing fear and concern among the population. Tumorigenicity refers to the propensity of cultivated cells to develop tumors. Two common in vitro tumorigenicity techniques are the Colony-forming assay and the Tumorsphere assay. The colony-forming assay, also known as clonogenic assay, assesses a single cell's ability to multiply and create a colony consisting of at least 50 cells [23]. This assay is typically utilized to differentiate the reproductive viability of untreated cells from cells treated with ionizing radiation or cytotoxic chemicals. It is also utilized to investigate the stemness and clonogenicity of stem cells.[24].

Colony forming assay is performed using the soft agar method. The basic steps involved in this assay are treating the cell monolayer in the flask, seeding the required number of cells on the agar and incubate for 1–3 weeks, fixing and staining the colonies and finally observing the colonies under the stereomicroscope [23]. Another in vitro tumorigenicity assay is tumorsphere assay which analyzes the potential of cancer stem cells (CSCs), a major cause of tumor initiation, progression, and recurrence after treatment. This assay is carried out under non-adherent conditions and serum-free medium supplied with growth factors of choice leading to the proliferation of CSCs and formation of spheres whereas the non-CSCs undergo apoptosis due to loss of adherence and abundant nutrients [25].

2.4 Cell invasion assay

Cell migration is a crucial biological process in which cells transform and move to a specific location within a suitable environment to carry out their function. It is a common biological process that occurs in almost all types of organisms. Changes or deregulation in cell migration or invasion patterns indicate pathogenic situations such as inflammatory disorders and cancer metastasis, with the latter being the most studied [21]. Several biological approaches used in the scientific community to analyze these occurrences include the cell culture wound-healing test, transwell migration and invasion assay, individual cell-tracking assay, and spreading assay. The assays attempt to offer pertinent data regarding the cell migratory pattern or its reaction to chemoattractants.

2.4.1 The cell culture wound-closure assay

It is the most basic way for determining the movement of

entire cell masses. Delving deeper, it can be utilized to analyze the morphological features and phenotypes of individual cells as they migrate. Tracking the proximity changes in relation to the control group at regular time periods reveals distinct migratory patterns or previously unidentified phenotypes. [26].

2.4.2 The transwell migration and invasion assay

The transwell migration and invasion assay assesses the capacity of individual cells to react to different chemoattractants such as chemokines, growth factors, lipids, or nucleotides. It also helps evaluate variations in cell movement caused by increased expression of a receptor. The study also identifies and describes the main regulators involved in cell migration. [26].

2.4.3 Individual cell-tracking assay or single-cell tracking assay

Performing single-cell tracking and live imaging under suitable settings enhances the benefits of a cell migration test. The program features a time-lapse video-microscopy procedure that analyzes cell populations with single-cell resolution through post-processing tracks. Understanding cell biology and lineage progression of different cell populations is really beneficial. [27].

2.4.4 Cell spreading assay

This experiment involves seeing and documenting the spreading process of individual cells using Differential Interference Contrast microscopy (DIC). The expanding state is captured every 5 seconds using a Charge-Coupled Device (CCD) camera, generating high-resolution grayscale images. Image capture may take several hours. [28].

2.5 Hybridoma technology and monoclonal antibodies

Antibodies, glycoproteins produced by immunoglobulins, are a key component of the immune system that B-cells manufacture to defend against infections. Antibodies are very specific and selective, making them valuable tools in bioengineering and biomedical research for an extended period. Antibodies are primarily categorized as Monoclonal Antibodies (mAbs) and Polyclonal Antibodies (pAbs) based on their origin from lymphocytes. Monoclonal antibodies are generated only by B lymphocytes or B cells and exhibit specificity for a single antigen. Because of this characteristic, monoclonal antibodies have a strong specificity and affinity for a single epitope of an antigen, while polyclonal antibodies are generated by various B-cells and have varying affinities for several epitopes of a certain antigen. Monoclonal antibodies (mAbs) are mass-produced by cultivating specialized cells that make antibodies, called 'Hybridomas', typically originating from mice, using a technique known as 'Hybridoma Technology' due to its great specificity. [29]. Hybridoma technology was invented by Georges Kohler and Cesar Milstein in 1975 and is regarded as a significant development. It has proved to be a robust, effective, and successful methodology employed in the field of biotechnology and biomedical research that solely deals with mAb isolation. B cells undergo antibody maturation in the germinal centers of secondary lymphoid organs such as lymph nodes, spleen, tonsils, and Peyer's patches. During proliferation, B cells undergo changes in the genes that encode the variable region of antibodies, which assist in selecting for high-affinity binding to the antigen. B cells produce antibodies that are formed by a natural combination of light chain and variable heavy chain genes together with constant region genes. Class Switch Recombination (CSR) is present in this location, unlike hybridoma technology where CSR is not seen. [29]. Following are the steps employed for the production of monoclonal antibody by hybridoma technology.

2.5.1 Isolation of antibody-producing B lymphocyte

The mice are immunized every 2–3 weeks with sheep red blood cells to generate B lymphocytes. The antibodies are obtained from the spleen cells of mice.

2.5.2 Screening of mouse for production of antibody

Following immunization, blood samples are collected from the mouse to measure the serum antibody titer. Once the titer reaches the ideal level, the mouse is given a boost by injecting antigen three days before fusing with myeloma cells [30].

2.5.3 Fusion of B cells with myeloma cells

When isolated spleen cells with a limited life span are fused with tumor lymphocytes, which are immortal, using PEG (Polyethylene Glycol), hybridomas with an endless life span are created.

2.5.4 Culturing of hybridomas

Hybridomas are cultured in a specialized medium that includes Hypoxanthine, Aminopterin, and Thymidine (HAT). Aminopterin in the medium inhibits the nucleotide synthesis route, causing cells to rely on an alternate pathway not found in myeloma cells.

2.5.5 Screening and selection of the desired colony

The cells are screened and chosen or selected for production of antibodies with the desired specificity.

2.5.6 Culturing of the selected hybridoma cells on large scale

The cells are cultured and used for the production of large quantities of antibodies [31].

2.5.7 Storage for future use

The cells are frozen and stored for future use in therapeutics.

2.6 Gene therapy

A gene is the basic biological unit of inheritance that consists of an organized sequence of nucleotides found in chromosomes. Genes act by encoding proteins or RNA molecules that determine inherited traits like hair and eye color and texture. Any changes in a gene sequence can result in aberrant gene function. Gene therapy is an experimental medical method that aims to correct non-functional or malfunctioning genes by replacing them with healthy and functional ones. Researchers have employed various methods to fix a faulty gene by replacing it with a healthy copy or by deactivating the mutant gene responsible for the disease. It has been extensively researched for a range of conditions including immunological deficiency, blood disorders, eye difficulties, metabolic disorders, nerve cell regeneration, and cancer [32]. The first case of gene therapy was discovered in the 1990s whereby a functional Adenosine Deaminase (ADA) gene was incorporated in the white blood cells of the patient, replacing the non-functional ADA [33]. This application yielded intriguing results regarding the immune systems, making it the most dependable strategy.

Gene therapy can be categorized into two basic methods: Ex-vivo gene therapy and In-vivo gene therapy. One involves transferring genes into patient cells outside the body, while the other involves transferring genes straight to cells inside the body. Various strategies are utilized for this purpose, such as direct or liposome-mediated DNA injection, calcium phosphate transfection, electroporation, dendrimers, hybrid methods, retrovirus, and other viral vectors. Gene therapy has been applied to the following clinical conditions:

2.6.1 Parkinson's disease (PD)

Gene therapy has been utilized to enhance the advanced symptoms of Parkinson's disease. Gene therapy was used to introduce the gene that produces 'Glutamic Acid Decarboxylase (GAD)' into the basal ganglia. Individuals with Generalized Anxiety Disorder (GAD) had elevated levels of the neurotransmitter Gamma-Aminobutyric Acid (GABA), which functions to suppress brain signals and reduce nervous system activity. Reduced GABA function causes specific brain diseases. [34].

2.6.2 Alzheimer's disease (AD)

AD and other frontotemporal dementias (FTDs) are caused by the accumulation of amyloid-β peptide (Aβ) and protein tau in the brain. It is characterized by having memory loss, difficulty in learning and communicating along with the inability to organize things. The use of recombinant Adeno-Associated Viruses (rAAVs) has provided new ways for studying AD and other related neurological disorders [35]. Such strategies or approaches have added novel dimensions to medical treatments

2.6.3 Cystic fibrosis

Cystic fibrosis primarily impacts the lungs. The symptoms include of inflammation, airway obstruction resulting in respiratory tract infection, and deformity. Introducing the CFTR gene directly into the epithelial cells of the respiratory tract can alleviate symptoms but does not provide a complete cure for those with cystic fibrosis. [36].

2.7 Cell therapy

Cell-based therapy is a crucial and well recognized method of treatment in the realms of modern science and medicine. Regenerative Medicine is a crucial practice in healthcare that aims to replace defective cells, tissues, or organs to restore their normal function, making it a significant treatment option for severe illnesses[4]. There has been a surge of interest and research in exploring the capabilities of stem cells over the years. They are cells that occur naturally in living organisms, known for their ability to self-renew indefinitely and to transform into specialized adult cell types. There are three primary categories of stem cells: Totipotent, Pluripotent, and Multipotent. [37].

Below is a description of numerous types of stem cells utilized in treating different illnesses:

2.7.1 Pluripotent stem cells

Allegedly, pluripotent cells have effectively treated animals. Animals with diabetes are treated with cells that have insulin receptors sensitive to glucose levels. The mice with acute spinal cord damage and vision impairment are treated using myelinated neurons and retinal epithelial cells, respectively. Researchers are currently studying the use of pluripotent stem cells to treat several ailments like Parkinson's disease, muscular dystrophy, and heart failure.

2.7.2 Induced pluripotent stem cells

Induced Pluripotent Stem Cells (iPSCs) are stem cells generated from normal adult somatic cells using gene and factor co-expression. These are crucial for preserving the distinctive traits of Embryonic Stem (ES) cells. iPSCs have been successfully utilized in treating several problems such as Parkinson's disease, spinal muscular atrophy, cardiac diseases, blood disorders, diabetes, amyotrophic lateral sclerosis, Huntington's disease, and familial dysautonomia, as reported.

2.7.3 Multipotent stem cells

In the 1960s, multipotent stem cells from bone marrow, known as Hematopoietic stem cells, were utilized to treat cancer such as leukemia, myeloma, and lymphoma. Mesenchymal stem cells capable of producing whole joints in mice models have been utilized for bone and cartilage regeneration. Treatments for cardiac conditions are now being tested in clinical trials.

Table 1. Merits and demerits of different 3D cell culture techniques.

Approach	Merits	Demerits
3D Spheroids	High reproducibility	Expensive to prepare
	Therapeutic inhibition to drug exposure can be easily evaluated through image analysis	Optimization protocols for each cell line is required
	Constant perfusion	Difficult light matter interactions for large spheroids
	Lower consumption of reagents	Differences in spheroids' diameters,
	Control of shear stress and pressure on cells	Low-throughput
	Capable of imbibing large amount of water or biological fluid	Labour intensity

Approach	Merits	Demerits
Hydrogels	Ease of maintenance	Low mechanical strength
	Amenable of controlling the micro-tissue size and large amount of micro-tissues per plate	Difficult to handle Expensive
Organoids	Amenable to high-throughput screening	Absence of microenvironment
	Long lived organoid production from single cells	Optimization protocols are not globally standardized.
		Organoid cultures rapidly die due to contamination
Cancer co-culture Models	Easily evaluate cell–cell interactions of cancer microenvironment	Microbial contamination
	Provide fluid flow	Static condition
	Easy to handle and quantify	
	Relevant mechanical cues	
Organ-on-a-chip	Enable stable co-culture of living human cells	Architectural complexity of developing human tissues and organs
	Good control over microenvironment	Difficulty in standardization and scale-up

3. APPLICATIONS OF 3D CULTURES IN VITRO

Recent advancements in cell culture technology have led to the rise of three-dimensional (3D) cell culture, which is enhancing tissue organization and cell differentiation by simulating sick tissue environments. An optimal 3D cell culture system consists of closely connected tissues facilitating cell-to-cell communication, resembling the dynamic extracellular matrix (ECM). It contains cell scaffolds in a fluid medium that promotes cellular differentiation (Table 1). An essential aspect of a 3D culture setting is the capacity to arrange cells spatially with neighboring cells while considering physical limitations. [8]. This significant approach has gardened great focus on understanding complex cellular biology and their responses by validating mammalian tissue studies via linking the gap between in vitro and in vivo environments. The two-dimensional (2D) cell cultures lack several features that 3D cultures impart such as tissue-specific architecture and complex cellular interactions that make them poor models for complex diseases. Based on the process of preparation, 3D techniques are categorized into (i) scaffold-based, (ii) scaffold-free culture systems. Scaffold-based technique efficiently is more responsive towards cell-to-ECM connection because of their potentiality of mimicking ECM whilst scaffold-free technique persuades physiological and cellular gradients. Scaffold or matrix is designed according to the tissue of interest, higher is the complexity of the scaffold, and the higher is the difficulty to extract the cells for analysis [9]. The scaffolds produced are primarily composed of polymeric hard material-based support, microfluidic-based assembly, hydrophilic glass fiber, and organoids. Scaffold-free systems create cellular aggregates known as spheroids. Techniques like magnetic levitation, liquid overlay, hanging drop microplates, and spheroid microplates have improved the formation of spheroids. However, the preference for 3D scaffolds has increased due to their ability to influence the behavior of cultured cells based on the matrix they are grown in.

3.1 3D Spheroids

3D spheroids, also known as multi-cellular aggregates, are small spherical cellular structures made up of various types of cells, typically created without the use of scaffolds. 3D spheroids exhibit key traits such as mimicking a cell's normal physiological functions, cellular diversity, gene activity, intercellular communication, and structural organization in relation to cell contact. [10]. Various types of 3Dspheroids include embryonic bodies, tumor spheroids (spheres of different tumorcells), hepatospheres (spheres of hepatic cells), neurospheres (spheres of different cell types of the central nervous system (CNS)) and mammospheres (spheres of mammary glands) [38]. An ideal 3D spheroid constitute ECM components such as proteoglycans, laminin, collagen, fibronectin, tenascin, and glycosaminoglycans [39] which tightens the spheroid density with close ECM-cell and cell–cell anchors eventually increase interstitial fluid pressure (IFP). Depending on the primary amount of cells seeded, the size of spheroid increases with an elevation in cell number, oxygen, and nutrient gradients equivalent to the tissue of interest [5]. Alongside, the different techniques enabling spheroid cultures are illustrated further. Hanging drop technique is a non-scaffold method wherein a drop of media containing cells are suspended inversely on the lid of the culture dish (bottom-less and open) such that there is no surface provided for the cells and tend to hang. This attempt forms a cluster called spheroid at the tip of the droplet when cultured for a longer period [40]. Spheroids formed through hanging drop cultures have fetched considerable stance in cell culture technology with 100% reproducibility owing to ubiquitous applications in cancer research [41], toxicity testing in hepatocytes [42], and constructing cardiac spheroids [43]. Another method involves the use of a liquid overlay that eases the formation of aggregates and commercially produced as low adhesion plates. These spheroid microplates contain either hydrophilic or hydrophobic coating with V-shaped bottom and allow mild attachment to the surface such that the cells tend to self-aggregate and form spheroid. Unlike the hanging drop technique, low adhesion plates generate one spheroid per plate that signifies its importance for multicellular culture. This ensures a medium-throughput screening that requires no modification in spheroid formation [44]. Spheroids can be cultivated using magnetic nanoparticles and a magnetic field. Magnetic cell levitation is a method commonly used to create spheroids of mesenchymal stem cells and for tissue engineering applications [45, 46]. An in vivo experiment demonstrated that human glioblastoma cells, when levitated using a magnetic technique, closely replicated the protein expression seen in human glioblastoma tumor xenografts. [47].

3.2 Organoids

Organoids are primary cultures created from cell aggregates in vitro, cultivated in 3D gels with extracellular matrix to form organ-like structures using physical support (cell adherence) or pharmacological cues (signaling pathway regulation). Embryonic, adult, primary, and stem cells are used in creating organoids. Organoids are categorized into tissue and stem cell organoids based on the organ-like structures they develop. The application of organoids has helped in producing numerous in vitro organoids such as rectal [48], gastric [49], lung [50], liver [51], pancreas [52], retina [53], thyroid [54], kidney [55] and intestine [56] that had successfully recapitulated the structural and functional motif of real organs. 3D organoids are extensively used by researchers to decipher the toxicity analysis, examine the genetic pathologies and investigate the local immune responses to infections. In addition to this, current reports have suggested the promiscuous application of organoids in platforms like transcriptomics and proteomics technologies. One such example illustrates the interaction study between Zika virus and Toll-like receptor 3 is performed by the generation of cerebral organoids from embryonic stem cells [57]. They have been also used as models for distinct genetic diseases. For instance, a study applied the rectal organoid model of cystic fibrosis for the investigation of the potency of transmembrane regulator-modulating compounds [48]. Besides, the tubular organoids model of polycystic kidney disease was also used to unravel the cause for cyst formation [55]. Apart from this, organoids have been an excellent source of models to understand the depth of neurodegenerative diseases viz.; Alzheimer, Parkinson's, HIV, diabetes, or cancer.

3.3 Cancer co-culture models

Cancer cell lines are now recognized as a prominent tool for understanding the intricate physiology of malignancies. Cell cultures have facilitated preclinical research by providing insights into disease processes, tissue morphology changes, gene function, cell biology, and tissue engineering. They have developed the ability to provide consistent samples without any alterations or differences. Monolayer cell cultures (2D) from solid tumors were unable to replicate the structural components of the tumor microenvironment. Recently, 3D cancer cell culture models have provided a lasting platform where the extracellular matrix (ECM) of the 3D structure mimics that of the original cell culture. This has enhanced our ability to forecast how tumors would respond to treatment. [59]. The application of 3D cell culture models of tumors have ought to manifest typical properties of tumor microenvironment such as gene and protein expressions, morphology, angiogenesis, malignancy and invasiveness. From this standpoint, 3D tumor cell culture models scintillate anticancer therapeutics and cancer drug discovery. To date, a vast content of literature owes the significance of these 3D co-cultures models in varying applications. In a study, tumor-associated macrophages (TAM) or cancer-associated fibroblasts (CAF) and gelatin hydrogel microspheres (GM) have been applied to produce cancer co-culture models from different cancer cells including HepG2 (liver), MCF-7 (breast) and WA-hT (lung) in order to inspect sustained release of drugs. They induced metastatic proteins involved in epithelial-mesenchymal transition (EMT) with transforming growth factor-β1 (TGF-β1) and reported elevation in N-cadherin and Vimentin proteins with deceleration in E-cadherin protein [58]. Recently, cancer co-culture models evinced interest in numerous approaches such as 3D breast cancer co-culture models obtained from MCF-7, MRC-5 and MDA-MB-231 tumor cells were used in investigating radiation-induced fibrosis [59], tumor-associated fibroblast differentiation [60] and development of immunotherapies [61], 3D lung cancer co-culture models derived lung squamous carcinoma and Non-Small Cell Lung Cancer Cells (NSCLC)fromTUM622, A549 and Colo699 tumor cells were utilized to explore tumor-stroma interactions [62, 63], 3D renal cancer co-culture models formed from Caki 1 (skin metastasis derived) and ACHN (pleural effusion derived) were sought for determining the efficacy of produced 3D models in stem cell physiology research and drug toxicity screening [64]. 3D colon cancer co-culture models acquired from LS 174 T, HCT 116, Colo205, MCF7, SW480, SW620, CCD-18Co, Caco-2, HT-29, and H446 have also been used to explore tumor-stroma interactions [65].

3.4 Tissue co-culture models

Utilizing co-culture cells in in vitro tissue models has shown increased versatility in replicating many systems of the human body, a challenging endeavor. These models have played a crucial role in replacing various animal models formerly used to study human physiology and pathology. Animal models are limited by their inability to accurately replicate key human characteristics and by ethical restrictions. This resulted in gaining insights into the development of tissue models using co-cultures as a research tool, aiming to establish models that mimic the natural microenvironment of cells and investigate the underlying causes of diseases. In vitro 3D tissue models have been successfully created by tissue engineering, exhibiting human traits with greater complexity than 2D monolayers. Some of the examples of 3D models constructed from tissue co-cultures include kidney [66], neuro-glia [67], lung [68], liver [69], ovary [70] and intestine [71] that have potentially predicted and represented physiological responses of the original culture of cells. Most probably, primary cells are used as they possess feasible in vivo features of not being immortal, incapable of getting transformed and consist of limited survival time in culture. Pertaining to these advantages, they have been applied to develop models that would combat various disease and physiological studies. Reportedly, tissue models have been revolutionized in terms of investigating multiple changes in real-time processes. A pulmonary endothelium model was constructed to investigate massive inflammation in patients with acute respiratory disease syndrome (ARDS). The authors performed this study using lavage samples of the patients for determining the etiology of ARDS that took place during the process of disease [72]. Endothelial cells have been crucial in building 3D tissue models due to their intriguing physiological functions. The authors were able to study epithelial cell interactions, microbiological infections, and cellular signaling by using in vitro gastrointestinal epithelial cell cultures from adult mouse colon [73]. A different example demonstrated the capability of hepatocyte tissue cultures in preserving the cancer cell hierarchy in human hepatocellular carcinoma [74].

3.5 Organ-on-a-chip

Organ-on-a-chip is a biomimetic technology that involves creating digital microchips and microfluids containing living cells to replicate the natural environment of certain organs. There are several factors that made organ chips be listed in "Top Ten Emerging Technologies" in the World Economic Forum [75] such as shear force, tissue-boundaries, concentration gradients, tissue–organ interactions and cell patterning. Organ chips have intensified in the field of drug therapeutics for their ability of high throughput screening. Table 2 summarizes the recent researches carried out using various organ chips. These organ chips use microtechnology that provides nutrients to the cells for their better growth and proliferation. Microfluids are one such component that has been used in various studies for efficient treatment in drug sensitivity testing [86]. Talking of this notion, a microfluidic chip was produced in order to monitor and document real-time impedimetric biosensor changes. Other organ-on-a-chip models such as blood–brain barrier chips have been developed to represent the in vivo architecture of brain involving micro blood vessels by using type 1 collagen hydrogel. Another model entails the significance of human-on-a-chip that depicts the normal human physiology in combination with single organ chip within a microfluidic system that ultimately forms multi-organ chip [87]. A study focused on an in vitro 3D-tumor-on-a-chip technology that demonstrated its significance in quorum sensing in tumor cells triggered by salmonella [88]. Organ-on-a-chip has become widely used in several research platforms because of its effective physiological bio-mimicry.

3.6 Patient-derived cells

Animal models in laboratories are largely avoided since they are expensive and demand a significant amount of labor. The method was substituted with in vitro models, which, despite numerous benefits, are limited by inadequate cell-to-cell and cell-to-matrix interactions. Cell cultures grown in a laboratory setting have been found to develop numerous genetic and epigenetic changes that lead to a loss of their original characteristics. These models have their own advantages and disadvantages in terms of cellular responsiveness, cellular makeup, and structural characteristics.

Table 2. Summary of recent organ-on-a-chip models.

Organ type	Incorporated cell types	Organ-specific properties	Ref.
Lung	Primary lung alveolar epithelial cells	Breathing movements	[76, 77]
	Primary lung endothelial cells	In vivo functionality array of tiny alveoli	
		Human lung parenchyma (lung alveoli and ultrathin air-blood barrier)	
		Recreates the native viscoelastic microenvironment of the cells	
	Human vascular endothelial cells	Alveolar capillary barrier in the human lung	[78]
	Human alveolar epithelial cells		
Skin	Peripheral perfusion fluid (PPF)	Franz diffusion cell system	[79]
		Drug absorption across the dermal barrier	
		Microfluidic Diffusion Chamber (MDC)	
Liver	Hepatic cell lines	Hepatoprotective effect assessment	[80]
		Hepatic activity (cell viability, albumin synthesis, urea secretion, and cytochrome P450 enzyme activities)	
		Drug Screening and Toxicity Testing	

Organ type	Incorporated cell types	Organ-specific properties	Ref.
Kidney	Human podocytes Glomerular endothelial cells	Human glomerular filtration barrier	[81]
		Functions and structure of the glomerulus	
	Human-derived renal proximal tubule epithelial cells	Cell Polarization-Dependent Cisplatin Toxicity	[82]
		Proximal tubule	
Heart	Cardiomyocytes	Contractile behavior (contraction force, frequency, and synchronization) of a 3D cardiac tissue construct Three-dimensional beating tissue from human cardiomyocytes	[83, 84]
Pancreas	Human pancreatic beta-cell line	Human pseudo-islets in biomechanical flow conditions	[85]

These models have both advantages and disadvantages in terms of cellular composition, replicating the original tumor architecture, tumor microenvironment (TME), and sensitivity to external stimuli. Thus, patient-derived cells are utilized in a co-culture-based method, cultivated in a culture medium enriched with various nutrients. 89]. Different studies have exemplified the use of patient-derived cells as a preclinical model in drug discovery (screening and responses) in several types of cells. Patient-derived xenograft (PDX) models are made of minute pieces of tumor tissue of the surgical patients, utilized for implantation into an immune-deficient mouse. In a study, Fong et al. used PDX models from prostate cancer to investigate tumor-stromal interactions via the use of a 3D hydrogel system [90]. Likewise, in another study, Liu et al. demonstrated the establishment of patient-derived cell cultures from colorectal cancer cells of biopsies of cancerous and non-cancerous tissue that could grow in in vitro culture indefinitely by recapitulating exactly the same phenotypic and genotypic features of the original tissue

[91]. Researchers have developed patient-derived 3D cultures utilizing scaffold-based organoid culture to maintain the genetic characteristics of the original tissue [92].

4. THREE-DIMENSIONAL CELL CULTURE IN DRUG DISCOVERY AND DEVELOPMENT

Drug discovery is a protracted and time-intensive procedure that involves multiple phases of testing and enhancement. This includes identifying the target, discovering leads, validating in pre-clinical stages, and conducting clinical trials. [3]. Due to the constant failure of drugs in Phase II and Phase III clinical trials, there has been constant pressure on the pharmaceutical industry to seek more novel drugs with lower side effects and cost-effectiveness. 3D cell culture has emerged as a significant high-throughput system that has uplifted the standards of cell culture [93]. Specifically, spheroids are considered the most reliable model for testing drugs in various diseases because of their capability of resembling the natural environment of original tissue [93]. The spatial organization of spheroids in different layers of cells leads to cellular death by forming reactive oxygen species [94]. In the case of investigating the effect in 3D spheroids, fluorescence microscopy plays a key role in determining pharmaceutical dispersion within spheroids (eg-doxorubicin and epirubicin) [95]. The capital importance of any drug testing involves cell-based assays that are efficient enough and easily reproducible compared to expensive animal models. Cell-based assays have shaped the physiological relevance of 2D cultures [96]. While the reaction may vary from technique to technique such as cell viability, proliferation, signaling and migration and drug to drug for achieving better sensitivity. It is now broadly accepted that compared to 2D cultures, 3D models serve the resemblance of the natural environment of original tissue efficiently and differently in 3D environments. Research has nested stance on novel 3D culture technologies that impart functional basis of tissues such as spheroids and organoids [97]. A study used 3D hydrogel-based model for the determination of drug sensitivity in HepG2 cell lines by comparing cytotoxicity effect with cytotoxicity (CT50) and lethal dose (LD50) values [98]. Organoid 3D models also aid as a resourceful tool for modeling neurodevelopmental disorders [98]. Microfluidic chips have also been utilized in drug sensitivity testing whereby a study elaborated its efficacy in lung cancer which was in combination with stromal cell lines [98]. ADMET evaluation of the medicine is generally conducted using in vitro cell culture procedures. 3D cell cultures have enhanced medication pharmacokinetics in multiple studies by utilizing different types of cell culture models.

5. CELL BASED MANUFACTURING OF THERAPEUTIC PROTEINS

The creation of therapeutic proteins utilizing human cell lines has significantly impacted various medical fields, such as biopharmaceutical research and vaccine manufacturing. Mammalian cell lines are ineffective for protein production because they often include post-translational modifications (PTMs) that mimic those found in human proteins. These cell lines exhibit exceptional selectivity in producing proteins similar to those normally generated in humans, offering an advantage over mammalian expression methods. [99]. One of the most routinely and high yields of proteins production is performed by using cell-based expression systems such as Chinese hamster ovary (CHO) a cell line that constitutes major advances such as accomplishment of gene amplification, specific productivity, better selection strategies, and devising greater expression units and advanced hosts. CHO cells have established their safety profile for 20 years from the production of its first recombinant biotherapeutic protein in 1986 [100]. Other human cell lines such as BHK-21 cells are used for the generation of few coagulation factors such as factor VIII [101]. There are two vital human cell lines namely, HEK293 and HT-1080 that are used to manufacture licensed products of human PTMs. The advancement in protein-based drug development and technologies has driven more towards the therapeutic proteins market that comprises of sales of these therapeutic proteins. The methods that are involved in the production of these proteins are pegylation, glycoengineering, albumin fusion, Fc-fusion, product purity, targeting, and functionality of therapeutic protein drugs. Few examples of therapeutic protein drugs which has been produced using protein engineering technologies and approved by the Food and Drug Administration (FDA) from the past five years are imiglucerase, Belimumab, alfa, coagulation factor IX recombinant human and albiglutide [102]. Sanofi, a French pharmaceutical business, enhanced its research and development strategy by acquiring Ablynx, a leading producer of unique therapeutic proteins, known for its nanobody technology platform.

6. CONCISE SUMMARY

Many research studies have shown that despite the existence of advanced organ-on-chip technologies and bioengineered 3D models, drug companies are hesitant to adopt them due to their innovative nature, which necessitates additional validation and characterization. Additionally, utilizing 3D cell culture models with high-throughput screening, combined with high-content analysis, allows for the discovery of clinically significant molecules. Challenges persist in using 3D cell cultures for drug discovery due to limitations in size, shape, complexity, and test protocols. Successful extraction of specific phenotypes for drug screening necessitates thorough standardization and optimization. Therefore, there are only a small number of 3D models that have limited accessibility because of restricted permeability. As protein treatments continue to advance, more advanced therapeutic protein products will be developed to enhance future research.

7. AT A GLANCE

Cell culture is a crucial in vitro technique used to study different aspects of cells or tissues from a living organism, including cell biology, tissue structure, disease mechanisms, cell signaling, drug effects, cancer research, and preclinical drug trials. Two primary types of cell cultures often employed are two-dimensional (2D) and three-dimensional (3D) cultures. The former has been utilized since the 1900s due to its simplicity and cost-effective maintenance as it creates a monolayer, while the latter is the more complex version and currently receiving the most attention. This chapter aims to elucidate the authentic meaning and importance for both civilizations. The discussion begins by clearly differentiating between the two and then explores their varied applications in vitro. 2D culture is important for studying cell motility and disease treatment through various assays and therapies. On the other hand, 3D culture involves models with spheroid structures made up of multiple cell layers, highlighting its application in drug discovery and development. The chapter ends with a detailed description of producing therapeutic proteins using cells.

ABBREVIATIONS

2D	Two-dimensional
3D	Three-dimensional culture
Aβ	Amyloid-β peptide
ADMET	Absorption, distribution, metabolism, excretion and toxicity
ADA	Adenosine Deaminase
AD	Alzheimer's disease
ARDS	Acute respiratory disease syndrome
CAF	Cancer-associated fibroblasts
CCD	Charge-Coupled Device
CFTR	Cystic fibrosis transmembrane regulator
CHO	Chinese hamster ovary
CNS	Central nervous system
CSCs	Cancer stem cells
CT50	Cytotoxicity 50 percent
CSR	Class Switch Recombination
DIC	Differential Interference Contrast microscopy

ECM	Extracellular matrix
ES	Embryonic Stem
EMT	Epithelial-mesenchymal transition
FTDs	Frontotemporal dementias
GABA	Gamma-Aminobutyric Acid
GAD	Glutamic Acid Decarboxylase
GM	Gelatin hydrogel microspheres
HAT	Hypoxanthine, Aminopterin, Thymidine
IFP	Interstitial fluid pressure
iPSCs	Induced pluripotent stem cells
LDH	Lactate dehydrogenase
LD50	Lethal dose 50 percent
MTT	3-(4,5-dimethythiazol2-yl)-2,5-diphenyl tetrazolium bromide
MTS	3-(4,5-dimethylthiazol-2-yl)-5-(3-carboxymethoxyphenyl)-2-(4-sulfophenyl)-2H-tetrazolium
MDC	Microfluidic Diffusion Chamber
mAbs	Monoclonal Antibodies
NAD	Nicotinamide adenine dinucleotide
NADH	Reduced nicotinamide adenine dinucleotide
NADPH	Nicotinamide adenine dinucleotide phosphate
NSCLC	Non-Small Cell Lung Cancer Cells
pAbs	Polyclonal Antibodies
PDX	Patient-derived xenograft
PD	Parkinson's Disease
PPF	Peripheral perfusion fluid
PTMs	Post-translational modifications
rAAVs	Adeno-Associated Viruses
SRB	Sulforhodamine B
TAM	Tumor-associated macrophages
TGF-β1	Transforming growth factor-β1
TME	Tumor microenvironment
XTT	2,3-Bis-(2-Methoxy-4-Nitro-5-Sulfophenyl)-2H-Tetrazolium-5-Carboxanilide.

REFERENCES

1. Hudu SA, Alshrari AS, Syahida A, Sekawi Z. Cell Culture, Technology: Enhancing the Culture of Diagnosing Human Diseases. J ClinDiagn Res. 2016;10(3):DE01-DDE5. DOI:10.7860/JCDR/2016/15837.7460
2. Verma A, Verma M, Singh A. Animal tissue culture principles and applications. Animal Biotechnology. 2020;269-293. DOI:10.1016/B978-0-12-811710-1.00012-4
3. Smith SC, Baras AS, Lee JK, Theodorescu D. The COXEN principle: translating signatures of in vitro chemosensitivity into tools for clinical outcome prediction and drug discovery in cancer. Cancer Res. 2010; 70,1753-1758. DOI: 10.1158/0008-5472.CAN-09-3562
4. Michelini E, Cevenini L, Mezzanotte L, Coppa A, Roda A. Cell-based assays: fuelling drug discovery. Anal Bioanal Chem. 2010; 398(1):227-238. DOI: 10.1007/s00216-010-3933-z
5. Kapałczyńska M, Kolenda T, Przybyła W, Zajączkowska M, Teresiak A, Filas V, Ibbs M, Bliźniak R, Łuczewski Ł, Lamperska K. 2D and 3D cell cultures - a comparison of different types of cancer cell cultures. Arch Med Sci. 2018;14(4):910-919. DOI:10.5114/aoms.2016.63743
6. Verma A, Verma M, Singh A. Animal tissue culture principles and applications. Animal Biotechnology. 2020; 269-293. DOI:10.1016/B978-0-12-811710-1.00012-4
7. Kamińska K, Szczylik C, Bielecka ZF, Bartnik E, Porta C, Lian F, Czarnecka AM. The role of the cell-cell interactions in cancer progression. Journal of cellular and molecular medicine. 2015; 19(2), 283-296. DOI: 10.1111/jcmm.12408
8. Ravi M, Paramesh V, Kaviya SR, Anuradha E, Solomon FDP. 3D cell culture systems: Advantages and applications. Journal of Cellular Physiology. 2015;230(1):16-26. DOI: 10.1002/jcp.24683

9. Fennema E, Rivron N, Rouwkema J, van Blitterswijk C, de Boer J. Spheroid culture as a tool for creating 3D complex tissues. Trends in Biotechnology. 2013;31(2):108-115. DOI: 10.1016/j.tibtech.2012

10. Zanoni M, Piccinini F, Arienti C, Zamagni A, Santi S, Polico R, Bevilacqua A, Tesei A. 3D tumor spheroid models for in vitro therapeutic screening: A systematic approach to enhance the biological relevance of data obtained. Sci. Rep. 2016;6:19103. DOI: 10.1038/srep19103

11. Duval K, Grover H, Han LH, Mou Y, Pegoraro AF, Fredberg J, Chen Z. Modeling Physiological Events in 2D vs. 3D Cell Culture. Physiology (Bethesda). 2017;32(4):266-277. DOI: 10.1152/physiol.00036.2016

12. Freimoser FM, Jakob CA, Aebi M, Tuor U. The MTT [3-(4,5-dimethylthiazol-2-yl)-2,5-diphenyltetrazolium bromide] assay is a fast and reliable method for colorimetric determination of fungal cell densities. Appl Environ Microbiol. 1999;65(8):3727-3729. DOI:10.1128/AEM.65.8.3727-3729.1999

13. Lomakina GY, Modestova YA, Ugarova NN. Bioluminescence assay for cell viability. Biochemistry (Mosc). 2015;80(6):701-713. DOI:10.1134/S0006297915060061

14. Kumar P, Nagarajan A, Uchil PD. Analysis of Cell Viability by the Lactate Dehydrogenase Assay. Cold Spring HarbProtoc. 2018; 2018(6). DOI:10.1101/pdb.prot095497

15. Khodabukus A, Prabhu N, Wang J, Bursac N. In Vitro Tissue-Engineered Skeletal Muscle Models for Studying Muscle Physiology and Disease. AdvHealthc Mater. 2018;7(15):e1701498. doi:10.1002/adhm.201701498

16. DeStefano JG, Jamieson JJ, Linville RM, Searson PC. Benchmarking in vitro tissue-engineered blood-brain barrier models. Fluids Barriers CNS. 2018; 15(1):32. DOI:10.1186/s12987-018-0117-2

17. Shen TW, Puccini B, Temnyk K, Herting S, Cardinal KO. Tissue-engineered aneurysm models for in vitro assessment of neurovascular devices. Neuroradiology. 2019; 61(6):723-732. DOI:10.1007/s00234-019-02197-x

18. Jakubowska W, Chabaud S, Saba I, Galbraith T, Berthod F, Bolduc S. Prevascularized Tissue-Engineered Human Vaginal Mucosa: In Vitro Optimization and In Vivo Validation. Tissue Eng Part A. 2020; 26(13-14):811-822. DOI:10.1089/ten.tea.2020.0036

19. Winkler J, Abisoye-Ogunniyan A, Metcalf KJ,Zena W. Concepts of extracellular matrix remodelling in tumour progression and metastasis. Nat Commun. 2020; 11(1), 5120. DOI:10.1038/s41467-020-18794-x

20. Grada A, Otero-Vinas M, Prieto-Castrillo F, Obagi Z, Falanga V. Research Techniques Made Simple: Analysis of Collective Cell Migration Using the Wound Healing Assay. J Invest Dermatol. 2017; 137(2):e11-e16. DOI:10.1016/j.jid.2016.11.020

21. Pijuan J, Barceló C, Moreno DF, Maiques O, Sisó P, Marti RM, Macià A, Panosa A. In vitro Cell Migration, Invasion, and Adhesion Assays: From Cell Imaging to Data Analysis. Frontiers in cell and developmental biology, 2019; 7, 107. DOI: 10.3389/fcell.2019.00107

22. Gau DM, Roy P. Single Cell Migration Assay Using Human Breast Cancer MDA-MB-231 Cell Line. Bio Protoc. 2020; 10(8):e3586. DOI:10.21769/bioprotoc.3586

23. Franken NA, Rodermond HM, Stap J, Haveman J, van Bree C. Clonogenic assay of cells in vitro. Nat Protoc. 2006; 1(5):2315-2319. DOI:10.1038/nprot.2006.339

24. Rajendran V, Jain MV. In Vitro Tumorigenic Assay: Colony Forming Assay for Cancer Stem Cells. Methods Mol Biol. 2018; 1692:89-95. DOI:10.1007/978-1-4939-7401-6_8

25. Wang H, Paczulla AM, Konantz M, Lengerke C. In Vitro Tumorigenic Assay: The Tumor Spheres Assay. Methods Mol Biol. 2018; 1692:77-87. DOI:10.1007/978-1-4939-7401-6_7

26. Justus CR, Leffler N, Ruiz-Echevarria M, Yang LV. In vitro cell migration and invasion assays. J Vis Exp. 2014; (88):51046. DOI:10.3791/51046

27. Gómez-Villafuertes R, Paniagua-Herranz L, Gascon S, de Agustín-Durán D, Ferreras M O, Gil-Redondo JC, Queipo MJ, Menendez-Mendez A, Pérez-Sen R, Delicado EG, Gualix J, Costa MR, Schroeder T, Miras-Portugal MT, Ortega F. Live Imaging Followed by Single Cell Tracking to Monitor Cell Biology and the Lineage Progression of Multiple Neural Populations. J Vis Exp. 2017; (130):56291. DOI:10.3791/56291

28. Fardin MA, Rossier OM, Rangamani P, Avigan PD, Gauthier NC, Vonnegut W, Mathur A, Hone J, Iyengar R, Sheetz MP. Cell spreading as a hydrodynamic process. Soft Matter. 2010; 6(19), 4788. DOI:10.1039/c0sm00252f

29. Parray HA, Shukla S, Samal S, Shrivastava T, Ahmed S, Sharma C, Kumar R. Hybridoma technology a versatile method for isolation of monoclonal antibodies, its applicability across species, limitations, advancement and future perspectives. IntImmunopharmacol. 2020; 85:106639. DOI:10.1016/j.intimp.2020.106639

30. Holzlöhner P, Hanack K. Generation of Murine Monoclonal Antibodies by Hybridoma Technology. J Vis Exp. 2017; (119):54832. DOI:10.3791/54832

31. Carvalho LS, Bravim da Silva O, Carneiro de Almeida G, Davies de Oliveira J, ParachinNS,Carmo TS. Production Processes for Monoclonal Antibodies. Intechopen; 2017. DOI:10.5772/61924(Chapter 10)

32. Gonçalves GAR, Paiva RMA. Gene therapy: advances, challenges and perspectives. Einstein (Sao Paulo). 2017;15(3):369-375. DOI:10.1590/S1679-45082017RB4024

33. Aiuti A, Roncarolo MG, Naldini L. Gene therapy for ADA-SCID, the first marketing approval of an ex vivo gene therapy in Europe: paving the road for the next generation of advanced therapy medicinal products. EMBOMol Med. 2017; 9(6):737-740. DOI:10.15252/emmm.201707573

34. Axelsen TM, Woldbye DPD. Gene Therapy for Parkinson's Disease, An Update. J Parkinsons Dis. 2018; 8(2):195-215. DOI:10.3233/JPD-181331

35. Combs B, Kneynsberg A, Kanaan NM. Gene Therapy Models of Alzheimer's Disease and Other Dementias. Methods Mol Biol. 2016; 1382:339-366. DOI:10.1007/978-1-4939-3271-9_25

36. Marquez Loza LI, Cooney AL, Dong Q, Randak CO, Rivella S, Sinn PL, McCray PB. Increased CFTR expression and function from an optimized lentiviral vector for cystic fibrosis gene therapy. MolTher Methods Clin Dev. 2021;21:94-106. DOI:10.1016/j.omtm.2021.02.020

37. Aly RM. Current state of stem cell-based therapies: an overview. Stem Cell Investig. 2020;7:8. DOI:10.21037/sci-2020-001

38. Edmondson R, Broglie JJ, Adcock AF, Yang L. Three-dimensional cell culture systems and their applications in drug discovery and cell-based biosensors. Assay Drug DevTechnol. 2014;12(4):207-218. DOI:10.1089/adt.2014.573

39. Yeon SE, No da Y, Lee SH, Nam SW, Oh IH, Lee J, Kuh HJ. Application of concave microwells to pancreatic tumor spheroids enabling anticancer drug evaluation in a clinically relevant drug resistance model. PLoS One. 2013;8(9):e73345. DOI:10.1371/journal.pone.0073345

40. Tung YC, Hsiao AY, Allen SG, Torisawa YS, Ho M, Takayama S. High-throughput 3D spheroid culture and drug testing using a 384 hanging drop array. Analyst. 2011;136(3):473-478. DOI: 10.1039/c0an00609b

41. Zhao L, Xiu J, Liu Y,Zhang T, Pan W, Zheng X, Zhang X. A 3D Printed Hanging Drop Dripper for Tumor Spheroids Analysis Without Recovery. Sci Rep 2019; 9, 19717. DOI: 10.1038/s41598-019-56241-0

42. Hurrell T, Ellero AA, Masso ZF, Cromarty AD. Characterization and reproducibility of HepG2 hanging drop spheroids toxicology in vitro. Toxicol In Vitro. 2018;50:86-94. DOI: 10.1016/j.tiv.2018.02.013

43. Bai Y, Yeung E, Lui C, Ong CS, Pitaktong I, Huang C, Inoue T, Matsushita H, Ma C, Hibino NA. Net Mold-based Method of Scaffold-free Three-Dimensional Cardiac Tissue Creation. J. Vis. Exp. 2018; 138, e58252. DOI:10.3791/58252

44. Raghavan S, Mehta P, Horst EN, Ward MR, Rowley KR, Mehta G. Comparative analysis of tumor spheroid generation techniques for differential in vitro drug toxicity. Oncotarget. 2016;7(13):16948-61. DOI: 10.18632/oncotarget.7659

45. Lewis EEL, Wheadon H, Lewis N, Yang J, Mullin M, Hursthouse A, Stirling D, Dalby MJ, Berry CC. A Quiescent, Regeneration-Responsive Tissue Engineered Mesenchymal Stem Cell Bone Marrow Niche Model via Magnetic Levitation. ACS Nano. 2016; 10, 9, 8346-8354. DOI:10.1021/acsnano.6b02841

46. Lewis NS, Lewis EE, Mullin M, Wheadon H, Dalby MJ, Berry CC. Magnetically levitated mesenchymal stem cell spheroids cultured with a collagen gel maintain phenotype and quiescence. J Tissue Eng. 2017; 8:2041731417704428. DOI: 10.1177/2041731417704428

47. Souza GR, Molina JR, Raphael RM, Ozawa MG, Stark DJ, Levin CS, Bronk LF, Ananta JS, Mandelin J, Georgescu MM, Bankson JA, Gelovani JG, Killian TC, Arap W, Pasqualini R. Three-dimensional tissue culture based on magnetic cell levitation. Nat Nanotechnol. 2010; (4):291-296. DOI: 10.1038/nnano.2010.23

48. Dekkers J, Berkers G, Kruisselbrink E, Vonk A, de Jonge H, Janssens HM, Bronsveld, I., van de Graaf EA, Nieuwenhuis EE, Houwen RH, Vleggaar FP, Escher JC, de Rijke YB, Majoor CJ, Heijerman HG, de Winter-de Groot KM, Clevers H, van der Ent CK, Beekman JM. Characterizing responses to CFTR-modulating drugs using rectal organoids derived from subjects with cystic fibrosis. SciTransl Med. 2016; 8:344ra84. DOI:10.1126/scitranslmed.aad8278

49. Seidlitz T, Koo BK,Stange DE. Gastric organoids—an in vitro model system for the study of gastric development and road to personalized medicine. Cell Death Differ. 2021; 28, 68-83. DOI: 10.1038/s41418-020-00662-2

50. Cunniff B, Druso JE, van der Velden JL. Lung organoids: advances in generation and 3D-visualization. HistochemCell Biol. 2021; 155, 301-308. DOI: org/10.1007/s00418-020-01955-w

51. Prior N, Inacio P, Huch M. Liver organoids: from basic research to therapeutic applications. Gut. 2019; 68(12):2228-2237. DOI:10.1136/gutjnl-2019-319256

52. Wills ES, Drenth JP. Building pancreatic organoids to aid drug development. Gut. 2017; 66(3):393-394. DOI:10.1136/gutjnl-2016-312865

53. Cowan CS, Renner M, De Gennaro M, Gross-Scherf B, Goldblum D, Hou Y, Munz M, Rodrigues TM, Krol J, Szikra T, Cuttat R, Waldt A, Papasaikas P, Diggelmann R, Patino-Alvarez CP, Galliker P, Spirig SE, Pavlinic D, Gerber-Hollbach N, Schuierer S, Aldin Srdanovic 8, Marton Balogh 3, Riccardo Panero 3, Akos Kusnyerik, Arnold Szabo A, Michael B Stadler MB, Selim Orgül S, Simone Picelli S, Pascal W Hasler PW, Andreas Hierlemann A, Scholl HPN, Roma G, Nigsch F, Roska B. Cell Types of the Human Retina and Its Organoids at Single-Cell Resolution. Cell. 2020; 182(6),1623-1640.e34. DOI: 10.1016/j.cell.2020.08.0

54. Saito Y, Onishi N, Takami, H, Seishima R, Inoue H, Hirata Y, Kameyama K, Tsuchihashi K, Sugihara E, Uchino S, Ito K, Kawakubo H, Takeuchi H, Kitagawa Y, Saya H, Nagano O. (). Development of a functional thyroid model based on an organoid culture system. Biochemical and biophysical research communications. 2018; 497(2), 783-789. DOI:10.1016/j.bbrc.2018.02.154

55. Cruz NM, Song X, Czerniecki SM, Gulieva RE, Churchill AJ, Kim YK, Winston K, Tran LM, Diaz MA, Fu H, Finn LS, Pei Y, Himmelfarb J, Freedman BS. (). Organoid cystogenesis reveals a critical role of microenvironment in human polycystic kidney disease. Nature materials. 2017; 16(11), 1112-1119. DOI: 10.1038/nmat4994

56. Bareiss PM, Metzger M, Sohn K, Rupp S, Frick JS, Autenrieth IB, Lang F, Schwarz H, Skutella T, Just L. Organotypical tissue cultures from adult murine colon as an in vitro model of intestinal mucosa. Histochemistry and cell biology. 2008; 129(6), 795-804. DOI: 10.1007/s00418-008-0405-z

57. Qian X, Nguyen HN, Jacob F, Song H, Ming GL. Using brain organoids to understand Zika virus-induced microcephaly. Development. 2017; 144(6):952-957. DOI:10.1242/dev.140707

58. Yakavets I, Francois A, Benoit A, Merlin JL, Bezdetnaya L, Guillaume Vogin G. Advanced co-culture 3D breast cancer model for investigation of fibrosis induced by external stimuli: optimization study. Sci Rep. 2020; 10,21273. DOI:10.1038/s41598-020-78087-7

59. Kunz-Schughart LA, Heyder P, Schroeder J, Knuechel R. A heterologous 3-D coculture model of breast tumor cells and fibroblasts to study tumor-associated fibroblast differentiation. ExpCell Res. 2001; 266(1):74-86. DOI:10.1006/excr.2001.5210

60. Saraiva DP, Matias AT, Braga S, Jacinto A, Cabral MG. Establishment of a 3D Co-culture With MDA-MB-231 Breast

Cancer Cell Line and Patient-Derived Immune Cells for Application in the Development of Immunotherapies. Front Oncol. 2020; 10:1543. DOI:10.3389/fonc.2020.01543

61. Chen S, Giannakou A, Golas J, Geles KG. Multidimensional Coculture System to Model Lung Squamous Carcinoma Progression. J. Vis. Exp. 2020; (157), e60644, DOI:10.3791/60644

62. Amann A, Zwierzina M, Gamerith G, Bitsche M, Huber J M, Vogel GF, Blumer M, Koeck S, Pechriggl EJ, Kelm JM, Hilbe W, Zwierzina H. Development of an Innovative 3D Cell Culture System to Study Tumour - Stroma Interactions in Non-Small Cell Lung Cancer Cells. PLoS ONE. 2014; 9(3), e92511. DOI:10.1371/journal.pone.0092511

63. Brodaczewska KK, Bielecka ZF, Maliszewska Olejniczak K, Szczylik C, Porta C, Bartnik E, Czarnecka AM. Metastatic renal cell carcinoma cells growing in 3D on poly-D-lysine or laminin present a stem-like phenotype and drug resistance. Oncol Rep. 2019;42(5):1878-1892. DOI:10.3892/or.2019.7321

64. Dolznig H, Rupp C, Puri C, Haslinger C, Schweifer N, Wieser E, Kerjaschki D, Garin-Chesa P. Modeling colon adenocarcinomas in vitro a 3D co-culture system induces cancer-relevant pathways upon tumor cell and stromal fibroblast interaction. Am J Pathol. 2011; 179(1):487-501. DOI:10.1016/j.ajpath.2011.03.015

65. Dolznig H, Rupp C, Puri C, Haslinger C, Schweifer N, Wieser E, Kerjaschki D, Garin-Chesa P. Modeling colon adenocarcinomas in vitro a 3D co-culture system induces cancer-relevant pathways upon tumor cell and stromal fibroblast interaction. The American journal of pathology. 2011; 179(1), 487-501. DOI:10.1016/j.ajpath.2011.03.015

66. Takasato M, Little MH. Making a Kidney Organoid Using the Directed Differentiation of Human Pluripotent Stem Cells. Methods Mol Biol. 2017; 1597:195-206. DOI:10.1007/978-1-4939-6949-4_14

67. Skaper SD, Facci L. Central Nervous System Neuron-Glia co-Culture Models and Application to Neuroprotective Agents. Methods Mol Biol. 2018; 1727:63-80. DOI:10.1007/978-1-4939-7571-6_5

68. Strikoudis A, Cieślak A, Loffredo L, Chen YW, Patel N, Saqi A, Lederer DJ, Snoeck HW. Organoids Derived from Human Pluripotent Stem Cells. Cell Rep. 2019; 27(12):3709-3723.e5. DOI:10.1016/j.celrep.2019.05.077

69. Coll M, Perea L, Boon R, Leite SB, Vallverdú J, Mannaerts I, Smout A, El Taghdouini, A, Blaya D, Rodrigo-Torres D, Graupera I, Aguilar-Bravo B, Chesne C, Najimi M, Sokal E, Lozano JJ, van Grunsven LA, Verfaillie CM, Sancho-Bru P. Generation of Hepatic Stellate Cells from Human Pluripotent Stem Cells Enables In Vitro Modeling of Liver Fibrosis. Cell Stem Cell. 2018; 23(1):101-113.e7. DOI:10.1016/j.stem.2018.05.027

70. Saadeldin IM, Kim SJ, Choi YB, Lee BC. Improvement of cloned embryos development by co-culturing with parthenotes: a possible role of exosomes/microvesicles for embryos paracrine communication. Cell Reprogram. 2014; 16(3):223-234. DOI:10.1089/cell.2014.0003

71. Jalili-Firoozinezhad S, Gazzaniga FS, Calamari EL, Camacho DM, Fadel CW, Bein A, Swenor B, Nestor B, Cronce MJ, Tovaglieri A, Levy O, Gregory KE, Breault DT, Cabral J, Kasper DL, Novak R, Ingber DE. A complex human gut microbiome cultured in an anaerobic intestine-on-a-chip. Nat Biomed Eng. 2019; 3(7):520-531. DOI:10.1038/s41551-019-0397-0

72. Murphy LS, Wickersham N, McNeil JB, Shaver CM, May AK, Bastarache JA, Ware LB. Endothelial glycocalyx degradation is more severe in patients with non-pulmonary sepsis compared to pulmonary sepsis and associates with risk of ARDS and other organ dysfunction. Ann Intensive Care. 2017; 7(1):102. DOI:10.1186/s13613-017-0325-y

73. Krump-Konvalinkova V, Bittinger F, Unger RE, Peters K, Lehr HA, Kirkpatrick CJ. Generation of human pulmonary microvascular endothelial cell lines. Lab Invest. 2001; 81(12):1717-1727. DOI:10.1038/labinvest.3780385

74. Kosuge M, Takizawa H, Maehashi H, Matsuura T, Matsufuji S. A comprehensive gene expression analysis of human hepatocellular carcinoma cell lines as components of a bioartificial liver using a radial flow bioreactor. Liver Int. 2007; 27(1):101-108. DOI:10.1111/j.1478-3231.2006.01410.x

75. World Economic Forum. Top Ten Emerging Technologies [Internet]. 2016. Available from:https://www.weforum.org/agenda/2016/06/top-10-emerging-technologies-2016/ [Accessed: 2016-06-23]

76. Stucki JD, Hobi N, Galimov A, Stucki AO, Schneider-Daum N, Lehr CM, Huwer H, Frick M, Funke-Chambour M, Geiser T, Guenat OT. Medium throughput breathing human primary cell alveolus-on-chip model. Scientific reports 8(1), 2018; 14359. DOI: 10.1038/s41598-018-32523-x

77. Zhang M, Xu C, Jiang L, Qin J. A 3D human lung-on-a-chip model for nanotoxicity testing. Toxicol Res (Camb). 2018;11;7(6):1048-1060. DOI: 10.1039/c8tx00156a

78. Zamprogno P, Wüthrich S, Achenbach S,Zamprogno P, Wüthrich S, Achenbach S, Thoma G, Stucki JD, Hobi N, Schneider-Daum N, Lehr CM, Huwer H, Geiser T, Schmid RA, Guenat OT. Second-generation lung-on-a-chip with an array of stretchable alveoli made with a biological membrane. Commun Biol. 2021;4(1):168. DOI:10.1038/s42003-021-01695-0

79. Lukács B., Bajza Á, Kocsis D, Csorba A, Antal I, Iván K, Laki A J, Erdő F. Skin-on-a-Chip Device for Ex Vivo Monitoring of Transdermal Delivery of Drugs-Design, Fabrication, and Testing. Pharmaceutics. 2019; 11(9), 445. DOI:10.3390/pharmaceutics11090445

80. Deng J, Cong Y, Han X, Wei W, Lu Y, Liu T, Zhao W, Lin B, Luo Y, Zhang X. A liver-on-a-chip for hepatoprotective activity assessment. Biomicrofluidics. 2020; 14(6):064107. DOI:10.1063/5.0024767

81. Petrosyan A, Cravedi P, Villani V, Angeletti A, Manrique J, Renieri AR, DeFilippo RE, Perin LP, Da Sacco SA. Glomerulus-on-a-chip to recapitulate the human glomerular filtration barrier. Nat Commun. 2019; 10, 3656. Doi:10.1038/s41467-019-11577-z

82. Nieskens TTG, Persson M, Kelly EJ, Sjögren AK. A Multicompartment Human Kidney Proximal Tubule-on-a-Chip Replicates Cell Polarization-Dependent Cisplatin Toxicity. Drug MetabDispos. 2020; 48(12):1303-1311. DOI:10.1124/dmd.120.000098

83. Sakamiya M, Fang Y, Mo X, Shen J, Zhang T. A heart-on-a-chip platform for online monitoring of contractile behavior via digital image processing and piezoelectric sensing technique. Med Eng Phys. 2020; 75:36-44. DOI:10.1016/j.medengphy.2019.10.001

84. Zhao Y, Rafatian N, Wang EY, Feric NT, Lai B, Knee-Walden EJ, Backx PH, Radisic M. Engineering microenvironment for human cardiac tissue assembly in heart-on-a-chip platform. Matrix Biol. 2020; 85-86:189-204. DOI:10.1016/j.matbio.2019.04.001

85. Zbinden A, Marzi J, Schlünder K, Probst C, Urbanczyk M, Black S, Brauchle EM, Layland SL, Kraushaar U, Duffy G, Schenke-Layland K, Loskill P. Non-invasive marker-independent high content analysis of a microphysiological human pancreas-on-a-chip model. Matrix Biol. 2020; 85-86:205-220. DOI:10.1016/j.matbio.2019.06.008

86. Zhiyun X, Yanghui G, YuanyuanH, Encheng L, Yan W, Jianing Z. Application of a microfluidic chip-based 3D co-culture to test drug sensitivity for individualized treatment of lung cancer. Biomaterials. 2013; 34:4109-17. DOI:10.1016/j.biomaterials.2013.02.045

87. Huh D, Hamilton GA, Ingber DE. From 3D cell culture to organs-on-chips. Trends Cell Biol. 2011;21(12):745-754. DOI:10.1016/j.tcb.2011.09.005

88. Swofford CA, Van Dessel N, Forbes NS. Quorum-sensing Salmonella selectively trigger protein expression within tumors. Proc. Natl. Acad. Sci. U S A. 2015; 112:3457-3462. DOI: 10.1073/pnas.1414558112

89. Ramzy GM, Koessler T, Ducrey E, McKee T, Ris F, Buchs N, Rubbia-Brandt L, Dietrich PY, Nowak-Sliwinska P. (2020). Patient-Derived In Vitro Models for Drug Discovery in Colorectal Carcinoma. Cancers (Basel). 2020; 12(6):1423. DOI:10.3390/cancers12061423

90.Fong EL, Wan X, Yang J, Morgado M, Mikos AG, Harrington DA, Navone NM, Farach-Carson MCA. A 3D in vitro model of patient-derived prostate cancer xenograft for controlled interrogation of in vivo tumor-stromal interactions. Biomaterials. 2016; 77:164-172. DOI:10.1016/j.biomaterials.2015.10.059

91. Liu X, Krawczyk E, A Suprynowicz F, Palechor-Ceron N, Yuan H, Dakic A, Simic V, Zheng YL, Sripadhan P, Chen C, Lu J, Hou TW, Choudhury S, Kallakury B, Tang DG, Darling T, Thangapazham R, Timofeeva O, Dritschilo A, Randell SH, ... Schlegel R. Conditional reprogramming and long-term expansion of normal and tumor cells from human biospecimens. Nat Protoc. 2017;12(2):439-451. doi:10.1038/nprot.2016.174

92. Beshiri ML, Tice CM, Tran C, Nguyen HM, Sowalsky AG, Agarwal S, Jansson KH, Yang Q, McGowen KM, Yin J, Alilin AN, Karzai FH, Dahut WL, Corey E, Kelly K. A PDX/Organoid Biobank of Advanced Prostate Cancers Captures Genomic and Phenotypic Heterogeneity for Disease Modeling and Therapeutic Screening. Clin Cancer Res. 2018; 24(17):4332-4345. DOI:10.1158/1078-0432.CCR-18-0409

93. Aparicio S, Hidalgo M, Kung AL. Examining the utility of patient-derived xenograft mouse models. Nat Rev Cancer. 2015;15(5):311-316. DOI:10.1038/nrc3944

94. Wartenberg M, Hoffmann E, Schwindt H, Grünheck F, Petros J, Arnold JRS, Hescheler J, Sauer H. Reactive oxygen species-linked regulation of the multidrug resistance transporter P-glycoprotein in Nox-1 overexpressing prostate tumor spheroids. FEBS Lett. 2005; 579(20):4541-4549. DOI:10.1016/j.febslet.2005.06.078

95. Gaspar VM, Costa EC, Queiroz JA, Pichon C, Sousa F, Correia IJ. Folate-targeted multifunctional amino acid-chitosan nanoparticles for improved cancer therapy. Pharm Res. 2015; 32(2):562-577. DOI:10.1007/s11095-014-1486-0

96.Turley SJ, Cremasco V, Astarita JL. Immunological hallmarks of stromal cells in the tumour microenvironment. Nat Rev Immunol. 2015;15(11):669-682. DOI:10.1038/nri3902

97. Liu F, Huang J, Ning B, Liu Z, Chen S, Zhao W. Drug Discovery via Human-Derived Stem Cell Organoids. Front Pharmacol. 2016;7:334. DOI:10.3389/fphar.2016.00334

98. Zhiyun X, Yanghui G, Yuanyuan H, Encheng L, Yan W, Jianing Z. Application of a microfluidic chip-based 3D co-culture to test drug sensitivity for individualized treatment of lung cancer. Biomaterials. 2013;34(16):4109-4117. DOI:10.1016/j.biomaterials.2013.02.045

99. Ghaderi D, Zhang M, Hurtado-Ziola N, Varki A. Production platforms for biotherapeutic glycoproteins. Occurrence, impact, and challenges of non-human sialylation. Biotechnol Genet Eng Rev. 2012; 28:147-175. DOI:10.5661/bger-28-147

100. KimJY, KimYG, LeeGM. CHO cells in biotechnology for production of recombinant proteins: current state and further potential. ApplMicrobiolBiotechnol. 2012;93(3):917-930. DOI:10.1007/s00253-011-3758-5

101. Wurm FM. Production of recombinant protein therapeutics in cultivated mammalian cells. Nat Biotechnol. 2004;22(11):1393-1398. DOI:10.1038/nbt1026

102. Dumont J, Euwart D, Mei B, Estes S, Kshirsagar R. Human cell lines for biopharmaceutical manufacturing: history, status, and future perspectives. Crit Rev Biotechnol. 2016;36(6):1110-1122. DOI:10.3109/07388551.2015.1084266

CHAPTER 4

ISOLATION AND EXPANSION OF MESENCHYMAL STEM/ STROMAL CELLS, FUNCTIONAL ASSAYS, AND LONG-TERM CULTURE ASSOCIATED ALTERATIONS IN CELLULAR PROPERTIES

1. INTRODUCTION: AN OVERVIEW

Mesenchymal stem/stromal cells (MSCs) are a type of multipotent stem/progenitor cell that were first identified in bone marrow by Friedenstein et al. around 50 years ago[1, 2]. They were characterized by their quick attachment to tissue culture capillaries and the formation of distinct "fibroblast" colonies. Julius Cohnheim, a German-Jewish pathologist, was the first to suggest over 150 years ago that a fibroblast-like cell population in bone marrow, which is not related to blood formation, played a role in wound healing[3]. Caplan first introduced the term "mesenchymal stem cell (MSC)" in the late 1980s[4]. MSCs have received significant attention in the past thirty years. Several research facilities specializing in Mesenchymal

Stem Cells (MSCs) have created several techniques for extracting and growing MSCs from different types of tissues. Assessing the properties of MSCs on various platforms and laboratories has highlighted the absence of a globally agreed criteria for defining MSCs. The Mesenchymal and Tissue Stem Cell Committee of the International Society for Cellular Therapy (ISCT) suggests three basic criteria to define MSCs[5]: adherence to plastic, expression of specific surface antigens, and ability to differentiate into three types of mesenchymal cells (osteoblasts, chondrocytes, and adipocytes).

Human MSCs from different tissues have the varied phenotypic features, the morphologic inconsistency, and heterogeneous functional behavior [6, 7, 8]. Indeed, the properties of stem cell have not been well established yet. Due to the unknown in vivo multipotent properties of MSCs, the issue of MSC nomenclature remains actively controversial. In 2019, ISCT MSC committee issued a position statement on nomenclature of MSCs clarifying the functional definition to emphasize the functional distinction of mesenchymal stem versus stromal cells [9]. MSCs have been considered as a promising therapeutic tool in tissue engineering and regenerative medicine. MSCs are well known to be present in almost every type of adult tissues, such as bone marrow [10, 11, 12], adipose tissue [10, 13, 14], lung [11, 15], synovial tissue [16, 17], dental pulp and periodontal ligament [18]. Notably, it has become apparent that MSCs are identified in the various human embryonic tissues, such as fetal bone marrow [19], fetal liver [20], aorta gonad-mesonephros and yolk sac [21], as well as multiple neonatal birth-associated tissues, such as placenta [10, 22, 23], amniotic and chorionic membrane [23, 24], umbilical cord tissue [10, 23, 24, 25], and umbilical cord blood [26, 27]. Therefore, different platforms/laboratories may use different type of tissue sources and methodologies for isolation and expansion of MSCs. This chapter firstly outlines protocols for standardized isolation and expansion of human bone marrow-derived MSCs (BM-MSCs), a major source of human MSCs, as well as BM-MSCs' characteristics, cryopreservation and thawing. Protocols for the preparation of MSCs derived from the other tissue types are similar to that of BM-MSCs, except tissue sample processing differentially. Human BM-MSCs are estimated at a very low frequency at approximately 0.001–0.01% of total nucleated cells [28, 29], and, therefore, human BM-MSCs are likely to be kind of difficult to isolate and harvest. This chapter will focus on optimal functional assays and their application based on our previous studies. It will be beneficial for researchers working with MSCs in basic research and translational and clinical applications, including osteogenesis, chondrogenesis, adipogenesis, CFU-F assay, 3-D cellular co-culture, MSC homing, and migration.

2. ISOLATION, CULTURE EXPANSION, PHENOTYPIC ANALYSIS, FREEZING AND THAWING OF HUMAN BM-MSCS

2.1 Materials

- Growth medium (pH 7.1–7.5): Dulbecco's modified Eagle's medium (DMEM)-low glucose (Sigma-Aldrich, St Louis, MO), 10% fetal bovine serum (FBS, heat-inactivated) (Atlanta Biologicals, Lawrenceville, GA), 0.37 g% sodium bicarbonate, and 1% penicillin–streptomycin. Adjust growth medium to pH 7.1–7.5.
- Ficoll-Paque (Density gradient medium) (Stem Cell Technologies, Cambridge, MA).
- 1 × Phosphate buffered saline (PBS), pH 7.2.
- 1 × PBS + 2% + 1 mM ethylenediaminetetraacetic acid (EDTA).
- Trypan blue, 0.4%.
- Freezing medium: 90% FBS and 10% (v/v) dimethyl sulphoxide (DMSO).
- Trypsin–EDTA solution 0.25% (Sigma, St. Louis, MO).
- RosetteSep™ Human Mesenchymal Stem Cell Enrichment Cocktail (Stem Cell Technologies, Cambridge, MA).

- Nalgene® Cryo 1°C Freezing Container.
- Cryogenic storage vials.
- Suitable box for storage in liquid nitrogen.
- Tissue culture ware: T25 flasks, T75 flasks, 100-mm tissue culture dishes, 6-well plates.
- Hemocytometer.
- Water bath at 37°C.
- Biological safety cabinet class II.
- Bench centrifuge with swinging bucket rotor.
- Inverted phase microscope.
- Sterile conical centrifuge tubes (15 and 50 mL).
- Sterile cell culture plastic pipettes (2, 5, 10, 25 mL).
- Automatic pipettor.
- 1.5 mL eppendorf tubes.

2.2 Isolation and culture expansion of human BM-MSCs

- Make sure BM sample, 1 × PBS + 2% FBS + 1 mM EDTA, Ficoll-Paque and centrifuge are all at room temperature.
- Spray the sample tube with 70% ethanol and conduct the isolation of BM-MSCs in the Biological Safety cabinet.
- For 10 mL size of BM, divide BM sample into two 50 mL conical tubes at room temperature. Add 50 µl RosetteSep Human Mesenchymal Stem Cell Enrichment Cocktail per mL of bone marrow and mix well.
- Incubate for 20 minutes at room temperature.
- Dilute sample with about same volume of 1 × PBS + 2% FBS + 1 mM EDTA solution; mix gently.
- Prepare two 50 mL conical tubes with 10 mL Ficoll-Paque each tube. Layer the diluted sample on top of Ficoll-Paque. Be careful to minimize mixing of Ficoll-Paque and sample. Tilt

the tube to 45 degree angle and slowly add the sample drop by drop to form a layer on top.

- Centrifuge at 1200 × g for 10 minutes at room temperature in a swinging bucket rotor and set the centrifuge to brake on.
- Remove the enriched cells from the Ficoll-Paque. Collect the interphase containing the mononuclear cells. Be sure not to touch the red pellet at the bottom. It is advisable to remove some of the Ficoll-Paque and a bit of upper plasma layer in order to endure complete recovery of the desired cells.
- Wash enriched cells with the 5× volume of 1 × PBS + 2% FBS + 1 mM EDTA solution. Spin 300 × g for 10 minutes.
- Carefully discard the supernatant and resuspend cells in 1 mL of MSC growth medium per tube. Perform a viable cell count with a hemocytometer using Trypan blue.
- Seed cells into one T-75 flask finally for 10 mL size marrow (a final cell concentration of 0.5–1.5 × 106 cells/cm2). 12 mL of MSC growth medium is supplemented.
- Put the flask in incubator at 37°C with 5% humidified CO2 for 48 hours to allow cells to attach.
- After 48 hours, observe with phase contrast microscopy and then remove growth medium and non-adherent cells.
- Wash cells twice with pre-warmed medium and add 13 mL of fresh MSC growth medium. Return the flask to the incubator.
- Change growth medium every 3 days and observe the cellular colony forming.
- CFU-F become in the next 3–5 days. Continue to culture until the cells reach 80% confluence in the 2 weeks.
- Remove the medium and wash with PBS 2–3 times.
- Add 3–4 mL pre-warmed trypsin–EDTA solution to cover cells in the flask. Return the flask to the incubator for 5 minutes.

- Check with phase contrast microscopy. When most cells become detached, gently tap the side of the flask.
- Add 5 mL growth medium to the flask to stop Trypsin–EDTA action. Resuspend cells by pipetting and transfer the entire cell suspension into a 15 mL conical tube.
- Centrifuge at 400 × g for 5 minutes.
- Remove the supernatant and resuspend the cells in 2–3 mL pre-warmed PBS. Centrifuge at 400 × g for 5 minutes.
- Repeat step 22.
- Harvest cells. This culture is considered as passage 0.
- Count cells and reseed cells at an optimum density of 5000 cells/cm2 in the appropriate tissue culture ware.
- After 24 hours, remove the growth medium and wash the cells attached to the plate once with PBS.
- Add the appropriate volume of fresh culture medium and incubate the cells for 2–3 days.
- When the culture reaches 80–90% confluence, remove the culture medium and wash with PBS 2–3 times.
- For trypsinization, add an appropriate volume of pre-warmed trypsin–EDTA solution to cover the entire cellular surface. Return the flask to the incubator for 5 minutes.
- Observe under a phase contrast microscope. When most cells become detached, gently tap the side of the flask.
- Add the appropriate volume of growth medium to stop Trypsin–EDTA action. Mix and collect the mixture of the entire cell suspension into a 15 mL conical tube.
- Centrifuge at 400 × g for 5 minutes at room temperature.
- Remove the supernatant without disturbing the cell pellet and resuspend the cells with an appropriate volume of pre-warmed PBS.

- Wash and centrifuge at 400 × g for 5 minutes again.
- Harvest cells. This culture is passage 1. MSCs at passage 1 can be frozen in liquid nitrogen (see the next) or continue to serially passage.

2.3 Characteristics of the expanded human BM-MSCs

Flow cytometry study of MSC characterisation utilizes the following antibodies: CD29-PE, CD34-PE, CD44-PE, CD73-PE, CD90-PE, CD45-FITC, CD147-FITC, HLA DR-FITC, IgG1-PE, and IgG1-FITC (BD Biosciences). IgG1 immunoglobulin serves as isotype negative controls, whereas passage 3 MSCs are identified using PE or FITC labeled antibodies targeting cellular surface markers.

- Remove the growth medium and wash cells with pre-warmed PBS twice.
- Add the pre-warmed trypsin–EDTA solution to the flask. Return the flask to the incubator for 5 minutes.
- When most cells become detached, gently tap the side of the flask.
- Add growth medium to the flask. Mix and transfer the entire cell suspension into a 15 mL conical tube.
- Centrifuge at 400 × g for 5 minutes.
- Remove the supernatant and resuspend the cells in the pre-warmed PBS.
- Centrifuge at 400 × g for 5 minutes.
- Carefully discard the supernatant and resuspend cells with pre-warmed PBS.
- Centrifuge at 400 × g for 5 minutes.
- Harvest cells.
- Count cells and make it a single cell suspension at concentration of 1x106 cells/mL in PBS.

- Add 250 µL cellular suspension to the appropriate number of FACS tubes.
- Add antibodies to the FACS tubes – 10 µL per antibody per tube.
- Incubate the FACS tubes for 20–25 minutes at room temperature on a shaker to prevent aggregation.
- Then spin down FACS tubes at 150 rpm for 5 minutes.
- Wash twice with PBS.
- Add 2 mL MSC growth media.
- Centrifuge the tubes at 150 rpm for 5 minutes.
- Pipette off the supernatant.
- Add 250 µL MSC growth media.
- Run samples through the FACS machine.

2.4 Freezing of BM-MSCs

Of note, perform all following steps under sterile conditions.

- Place freezing medium on ice.
- Harvest the cultivated MSCs, as described above (steps 1–10 in Section 2.3).
- Resuspend MSCs in freezing medium at a concentration of 5–10 × 105 cells/mL.
- Transfer cells into appropriate cryogenic storage vials or tubes and close the lid.
- Place the vials in the pre-cooled Cryo 1°C Freezing Container quickly and store at -80°C directly.
- After 24 hours, transfer the cryogenic storage vials to the suitable boxes to liquid nitrogen for long-term storage.

2.5 Thawing of BM-MSCs

Perform all following steps under sterile conditions.

- Place the frozen vial of MSCs rapidly into a 37°C water bath.
- Gently swirl the cryovial until the ice in the vial has melted.
- After thawing, transfer the entire content of the cryovial into a sterile 15 mL conical tube containing 5 mL of pre-warmed PBS.
- Gently swirl and centrifuge at 400 × g for 5 minutes.
- Aspirate supernatant completely.
- Wash again.
- Count and reseed cells at a density of 5000 cells/cm2 in the appropriate tissue culture ware.

3. FUNCTIONAL ANALYSIS OF EXPANDED HUMAN MSCS

3.1 In vitro osteogenesis of human BM-MSCs

Osteogenic medium: MSC growth medium supplemented with 50 μM ascorbic acid phosphate (AsAP) (Wako Chemicals USA, Richmond, VA), 0.1 μM dexamethasone (Sigma-Aldrich, St Louis, MO), and 10 mM β-glycerol phosphate (Sigma-Aldrich, St Louis, MO).

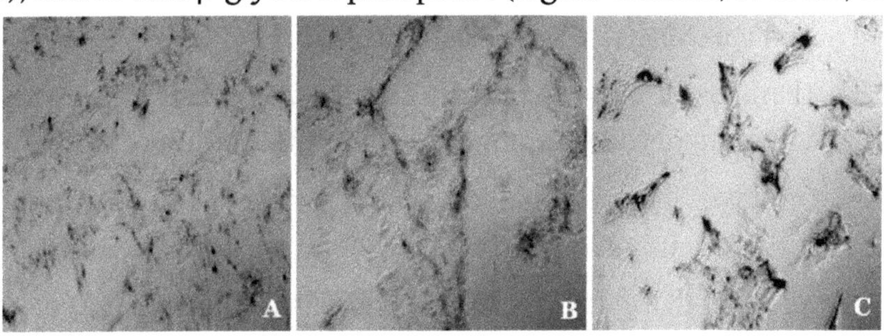

Figure 1. Osteogenic culture of human BM-MSCs. Representative images (10×) of human BM-MSC osteogenesis on day 12 (A), day 17 (B), and day 26 during osteogenic culture.

Several osteogenic markers, such as alkaline phosphatase (ALP), leptin receptor, and cathepsin K, are used as the indicator of early osteogenesis [28], calcium deposition as the indicator of late-stage osteogenesis [28, 29]. ALP activity and calcium deposition are exemplified to assess osteogenic differentiation of human MSCs from passage 4 (Figure 1).

3.1.1 Colorimetric quantitative ALP activity

- Collect cells from passage 4.
- Prepare cellular suspension in growth medium.
- Plate cellular suspension in 6-well plates at 1 × 105 cells/well in triplicate.
- After 24 hours, remove the growth medium and wash cells with osteogenic medium once.
- Add 2 mL osteogenic medium in each well.
- Culture cells in incubator at 37°C with 5% humidified CO_2.
- Change osteogenic medium every 3 days.
- Measure ALP activity about 1–2 weeks after MSC osteogenetic culture [29, 30]. Aspirate medium and gently wash cells twice with PBS.
- Add 600 μL of lysis buffer (0.5% Triton-X 100 in molecular grade ddH_2O) to each well and then scrape the cells off the surface using the end of a pipette tip.
- Transfer lysates to centrifuge tubes. Prepare 15 mL conical tubes equal to the number of tubes maintaining lysates.
- Dissolve the contents of a 40 mg capsule of phosphatase substrate (Sigma) in 10 mL ddH_2O. Scale up if necessary.
- Add 500 μL of 1.5 M alkaline buffer solution (Sigma) to each tube. Add 500 μL of phosphatase substrate solution into each tube. Keep tubes in 37°C water bath.
- Vortex samples and add 100 μL of each lysate to 15 mL tubes within 30 seconds. Fifteen minutes after the first sample is added, add 1 mL of 1 N NaOH to each tube in 30 second intervals removing tubes from water bath. The reaction will take place for 15 minutes at 37°C for each tube.

- Prepare standard curve (Sigma) (Note: prepare 100 nmol/mL solution of p-nitrophenol by combining 100 μL of 10 mM p-nitrophenol standard solution with 9.9 mL 0.02 N NaOH).
- Add 300 μL of standards and samples in triplicate to a 96-well plate.
- Measure the absorbance using excitation filter of 405 nm. One enzyme unit of ALP is defined as the quantity of enzyme which produces 1 nmol p-nitrophenol per 15 minutes [29].

3.1.2 Colorimetric quantitative calcium assay

- Aspirate or pipette out all culture medium from each well of the 6-well culture plate that contains induced or control cells to be tested.
- Rinse the cells twice with PBS. Add 200 μL of PBS to the side of each well, not to dislodge the cells. Aspirate off the PBS and re-rinse.
- Add 125 μL of 0.5 N HCl to each well.
- Scrape the cells off of the surface using a cell scraper and transfer the cells and HCl to a 1.5 mL polypropylene microcentrifuge tube with a tight fitting cap.
- Add an additional 125 μL of 0.5 N HCl to each cell to recover any cells remaining in the well, and transfer this to the appropriate tube.
- Samples may be capped tightly and stored at -20°C if they are not to be tested immediately.
- Extract the calcium from the cells by shaking the tubes on an orbital shaker for 4 hours at 4°C. If using frozen samples, allow extra time for samples to thaw.
- Centrifuge the sample tubes at 500 g for 2 minutes.
- Carefully collect the supernatant with extracted calcium, without disrupting the cell pellets, and transfer these to a new tube.

- Prepare a standard curve with the calcium standard and determine the amount of calcium in each control and osteogenesis-induced samples. Follow the instructions provided in the Stanbio Total Calcium LiquiColor® Procedure No. 0150 (Stanbio Laboratory).
- Three μL of sample vs. 297 μL of assay reagent (1:100 ratio for sample to reagent) for 96-well plate is used for each calcium determination. Assay reagent is mixed by equal volume of two solutions (Color Reagent and Base Reagent) provided in the kit. Distribute the assay reagent by multipipettor after adding samples. Absorbance is read at 550 nm.
- Unused sample extract may be re-frozen for future re-assay. If the reading was out of range, the sample can be diluted with ddH2O in a total volume of 3μL (e.g. 1 μL of sample+2 μL of ddH2O) and re-assayed again.

3.1.3 3-D culture of human BM-MSCs and calcium deposition measured by von Kossa staining

PuraMatrix™ hydrogel (BD Biosciences) is a 16-amino acid synthetic peptide hydrogel composed of a repeating sequence of arginine, alanine, aspartate, and alanine (RADA16) [31], which is widely used for 3-D culture.

- 3-D culture of human BM-MSCs on PuraMatrix hydrogel as follows.
- The stock of 1% peptide solution can be sonicated for 30 minutes to decrease its viscosity and then diluted with sterile ddH2O to a final concentration of 0.25% (w/v).
- 300 μL of 0.25% gel solution is loaded into each well of the 24-well plate until it is uniformly spread.
- The gelation is initiated by slowly dripping the medium along the wall of the well. 300 μL of medium is again added carefully on top and the plate is incubated at room temperature for one hour equilibration.

- After the peptide has assembled into hydrogel, the medium is changed two times over one hour to equilibrate the growth environment to physiological pH.
- The equilibrated samples are stored overnight at 37°C incubator and the cells are seeded the next day.
- Prepare the total of 4 × 104 cells suspended in MSC growth medium and 4 × 104 cells are seeded onto the hydrogel. The following day (Day 0), the medium will be replaced by osteogenic medium.
- Von Kossa staining can be conducted at day 24 after differentiation [30]. MSCs are rinsed with the Tyrode's balanced salt solution and fixed with 10% buffered formalin (Fisher Scientific) for 30 minutes.
- Incubated with 2% silver nitrate solution for 10 minutes in the dark.
- Rinse with ddH2O and expose to light for 15 minutes.
- Bright-field images of stained samples are captured with an inverted microscope.

3.1.4 3-D cellular culture conducted by encapsulation of human MSCs in PuraMatrix™ hydrogel

- To generate a 0.25% final concentration of PuraMatrix™ hydrogel for cells encapsulation, one part of 1% PuraMatrix™ hydrogel is diluted with same volume of sterile 20% sucrose, to reach 0.5% PuraMatrix™ hydrogel in 10% sucrose, and then mix with one part of 2× concentration of cells resuspended in 10% sucrose.
- For 24-well plates, 300 μL of PuraMatrix mixture is loaded into each well and 300 μL of medium is layered on top of the gel. The gelation of the PuraMatrix is completed in an incubator for 60 minutes.
- Change medium the next day.

- Von Kossa staining can also be conducted, as described above, or collect cells as follows for other experiments.
- Mechanically disrupt BD PureMatrix™ and cells in the well or cell culture insert by pipetting the media and gel up and down.
- Transfer to a 15 mL conical tube.
- Rinse out the well or cell culture insert using PBS.
- Centrifuge at 150 × g for 5 minutes. Discard supernatant. The pellet at the bottom of the tube contains cells and BD PuraMatrix fragments.
- Resuspend pellet in 2 mL of PBS. Spin and collect pellet again.
- Resuspend pellet in 1 mL of trypsin–EDTA and incubate at 37°C for 5–10 minutes. This will help separate cells that are still attached to each other.
- Add 5 mL PBS to spin cell pellet again.
- Aspirate the supernatant (do not disturb the gel). Resuspend pellet again in 1 mL of trypsin–EDTA and incubate at 37°C for 5–10 minutes.
- Add 5 mL PBS to spin cell pellet again.
- Aspirate the supernatant. Carefully take out one third of the gel pellet. Do not disturb the bottom (two-third) part of the gel.
- Wash with PBS twice.
- Add appropriate amount of lysis buffer to perform cell lysis and collect cell sample.

3.2 3-D cell pellet culture and chondrogenesis of human BM-MSCs

Chondrogenic medium: 95% DMEM high-glucose medium (Sigma-Aldrich, St Louis, MO), 1% 1 × ITS+1 solution (BD Biosciences, San Jose, CA), 1% Pen-Strep,

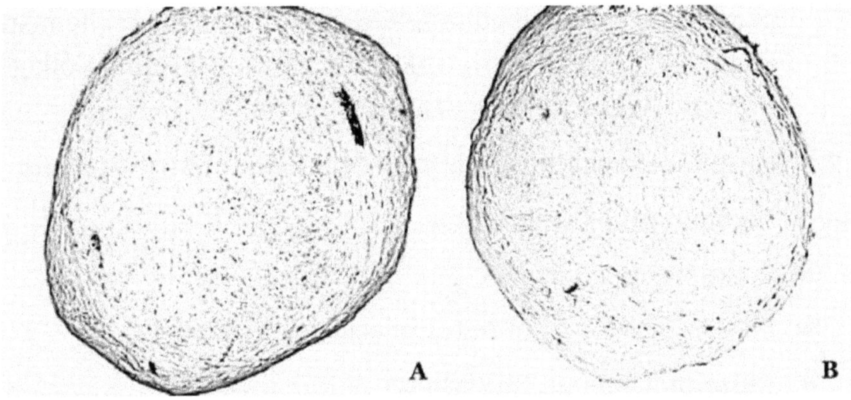

Figure 2. Chondrogenic culture of human BM-MSCs. Representative images (10×) of immunohistochemical staining of collage II (A) and aggrecan (B) in pellet culture samples at day 17 during chondrogenesis [29].

100 µg/mL sodium pyruvate (Invitrogen, Carlsbad, CA), 50 µg/mL AsAP, 40 µg/mL L-proline (Sigma-Aldrich, St Louis, MO), 0.1 µM dexamethasone, and 10 ng/mL recombinant human TGF-β3 (Lonza, Walkersville, MD).

Passage 4 human BM-MSCs are used for chondrogenic differentiation. Chondrogenic differentiation is induced by chondrogenic medium (Figure 2).

- In order to form a chondrogenic pellet, approximately 2.5 × 105 human BM-MSCs are centrifuged down in a 15 mL conical tube at 150 × g for 5 minutes at room temperature.
- Five hundred µL of chondrogenic medium is used to resuspend the 2.5 × 105 cells to a final concentration of 0.5 × 106 cells/mL.
- Cells are centrifuged down again. Place the 15 mL conical tube in the incubator at 37°C with 5% humidified CO2.
- MSCs are shaped into the pellet after 24 hours incubation without disturbance.
- The cell pellet is fed every 3 days for about 2–4 weeks and, after that, the chondrogenic pellet is harvested and sample

processing is described as follows for immunohistochemistry analysis (e.g. examination of the expression levels of Collagen II, X and Aggrecan) [29, 32].

3.2.1 Sample processing for immunohistochemistry analysis

Day 1: Fixation and dehydration

- Rinse specimen in PBS.
- Fix samples for 45 minutes in acid-formalin at 4°C.
- Rinse specimen in PBS twice.
- Embed specimen in 2% agarose.
- Transfer to vial and treat with 50% absolute ethanol for 1 hour at room temperature.
- Transfer to 70% Absolute ethanol for 1 hour at room temperature.
- Transfer to 95% Absolute ethanol for 1 hour at room temperature.
- Repeat the step 7.
- Transfer to 100% absolute ethanol for 1 hour at room temperature.
- Leave specimen at 4°C overnight.

Day 2: Clearing and infiltration

- Transfer to 100% CitriSolv (Fisher Scientific, catlog 22-143-975) for 1 hour at room temperature.
- Transfer to 100% CitriSolv for 1 hour at 55°C.
- Transfer to 1:1 mixture of CitriSolv/Micro-cut paraffin for 1 hour at 55°C.
- Repeat the step 3.
- Transfer to 100% Micro-cut paraffin for 1 hour at 55°C.
- Repeat the step 5.

Day 3: Embedding

- Transfer to 100% Micro-cut paraffin for 1–2 hours at 55°C.
- Position specimen in Peel-Away mold with 100% paraffin.
- Allow specimen to harden overnight.
- Specimen may be sectioned the following day.
- Conduct immunohistochemistry.

3.3 In vitro adipogenesis of human MSCs

Adipogenic medium: DMEM (1 g/L glucose), 10% FBS, 1% penicillin/streptomycin, 10 mg/mL insulin, 1 mM dexamethasone, 0.5 mM methylxanthine, and 200 mM indomethacin.

Passage 4 human BM-MSCs are used for adipogenic differentiation. Adipogenic differentiation is induced by adipogenic medium.

- Harvest cells from passage 4, as described in the previous section.
- Resuspend cells in adipogenic medium carefully.
- Transfer the single cell suspension in triplicate to 6-well plates (1 × 105 cells/well).
- Culture cells in the incubator at 37°C with 5% humidified CO2 for 3 weeks.
- Change medium every 2–3 days.
- After 3 weeks, aspirate adipogenic medium and wash cells twice with PBS.
- Cells are fixed in ice-cold methanol for 10 minutes.
- Aspirate methanol completely and wash cells twice with ddH2O.
- Add Oil Red O staining reagent to the wells to stain lipid vacuoles at room temperature and mix slowly about 20 minutes on a shaker plate.
- Aspirate Oil Red O staining reagent and wash cells twice with ddH2O.

- Observe and check the stained cells with phase contrast microscopy.

3.4 CFU-F assay

CFU-F assay can be used in vitro to evaluate the proliferation potential of MSCs. CFU-F assay is a well-established method for the quality control of MSCs' preparation. This section describes a traditional assay for CFU-F to evaluate the colony forming ability of human MSCs.

- Collect cells from passage 4.
- Prepare the single cell suspension in growth medium and seed cells in the 6-well plates in triplicate at three different densities, 1.5 × 105, 2.5 × 105, and 5 × 105 cells/well, in 2 mL growth medium, respectively.
- Culture cells in the incubator at 37°C with 5% humidified CO2 for two weeks.
- Change medium twice each week and check with phase contrast microscopy daily to prevent overgrowth. Stop cell culture as soon as colonies are forming visibly and proceed with the Giemsa staining of CFU-F colonies on a benchtop as follows.
- Wash the culture dishes twice with PBS.
- Fix cells by adding 2 mL methanol to each well for 5 minutes at room temperature.
- Gently remove the methanol and discard into the bio-hazardous waste.
- Air dry the culture vessels and add the diluted Giemsa staining solution for 5–10 minute at room temperature.
- Remove Giemsa staining solution and wash twice with ddH2O.
- Count visible colonies manually with a diameter greater than 5 mm.

3.5 Co-culture of MSCs with cancer cell line

There are different 2-D or 3-D models involving human MSCs and other cell sources, with or without dyes, used to investigate cell-cell contact, cell proliferation, MSCs' immunomodulatory abilities, and the specific cellular roles of each cell type.

These methods making co-cultures of MSCs and other cells are well-established, such as co-cultures of MSCs and human peripheral blood mononuclear cells [33, 34], MSCs and T cells [35], MSCs and human hematopoietic stem cells [36], MSCs and umbilical vein endothelial cells [37]. MSC-cancer cell (PC-9) co-culture will be described in this section (Figure 3).

3.5.1 Preparation of MSCs with CellTracker™ Red dye (CellTracker™ CM-Dil, C7000)

- Culture MSCs in an appropriate tissue culture ware (e.g. a 6-well plate)
- Remove culture medium and wash with PBS twice.
- Prepare CellTracker™ working solution in MSC growth medium. Make stock solution of lipophilic tracers in DMSO at 2 mM. Dilute the stock solution in MSC growth medium at 2 μM.
- Add the working solution in the tissue culture vessel to cover the entire cell surface.
- Incubate for 5 minutes or less in incubator at 37°C with 5% humidified CO2.
- Remove the CellTracker™ working solution.

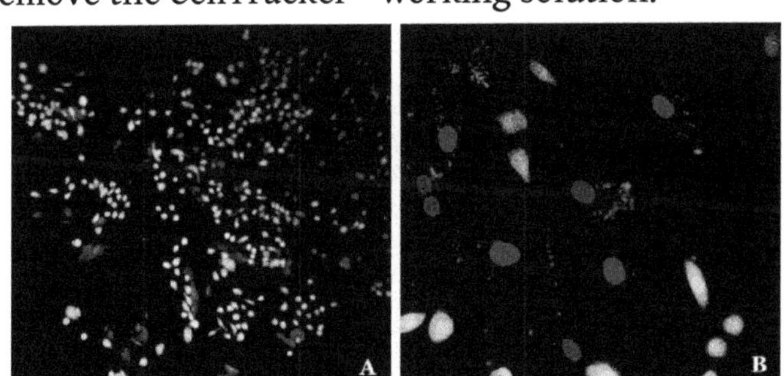

Figure 3. Human BM-MSCs-PC-9 co-culture. Representative images ((A) 5×; (B) 20×) of co-cultures of human BM-MSCs labeled with CellTracker™ red dye and PC-9 cells transfected with copGFP, nuclei counterstained with DAPI.

- Wash with MSC growth medium once.
- Add MSC growth medium and return the tissue culture vessel to the incubator.

3.5.2 Preparation of PC-9 cells labeling with GFP

- PC-9 cells labeling with copGFP (Santa Cruz, sc-108083) is performed using the lentiviral technique. Lentivirus is produced by using Lipofectamine™ 2000 (Invitrogen), according to the manufacturer's instructions. After transfection, PC-9 cells expressing copGFP may be isolated via puromycin (2 μg/mL) selection.
- Continue to serially passage.
- Collect PC-9 cells for cell co-culture.

3.5.3 MSC-PC-9 cell co-culture

- Harvest MSCs with CellTrcker™ red dye and wash with PBS once.
- Prepare the single cell suspension in growth medium at an appropriate cell concentration and transfer the cell suspension in a 6-well plate or a special chamber.
- Culture cells in the incubator at 37°C with 5% humidified CO2 for about 3 hours to allow cells to attach.
- Change fresh growth medium slowly.
- Add PC-9 cells in the 6-well plate (the same cell number of MSCs). Gently tap the side of the flask.
- Return the 6-well plate to the incubator and culture cell overnight.
- Check the cell co-culture under the contrast microscope and image under microscopy.

3.6 In vitro migration assay

Intercellular communication can occur through direct cell-to-cell interaction or paracrine transmission involving active substances.

Key signaling molecules such as cytokines, growth factors, and chemokines are produced by several cell types, including tumor cells and MSCs, in reaction to signals like inflammation or the tumor microenvironment. Signaling molecules drive circulating MSCs to home and migrate to locations of tissue injury or illness. MSCs must be able to move and locate damage sites in order to facilitate tissue repair and regeneration. There is a shortage of clinical data about the homing and distribution of transplanted MSCs throughout the body, despite several in vivo research being carried out on the homing and migration paths of MSCs for specific stem cell-based treatments. Various strategies exist for enhancing MSC homing and migration. This section assesses the in vitro migratory ability of human BM-MSCs using 8 µm-pore transwell chamber inserts from Corning.

- Harvest MSCs and prepare cell suspension in the serum-free medium.
- Transfer the cell suspension to the upper layer of a transwell insert at a density of 4 × 104 cells/cm2 and allow cells to migrate to the lower compartment containing MSC growth medium overnight in the incubator at 37°C with 5% humidifies CO2.
- Cells from the upper chamber of transwell are migrated. Gently scrape the MSCs using the cotton swab at the upper layer of the membrane.
- The migrated MSCs at the lower layer are stained with 0.1% crystal violet.
- Check the cells and image under a light microscope.
- Count the number of stained MSCs manually.

4. IN VITRO LONG-TERM CULTURE ASSOCIATED ALTERATIONS OF MSCS

4.1 Morphological and immunophenotypic alterations of MSCs

It is well known that MSCs demonstrate biological alterations in the course of in vitro long-term culture. Different tissue derived MSCs may present different morphological and immunophenotypic characteristics in the expansion culture. At present, there is lack of a unifying definition for the "passage" of MSCs. Morphological changes are continuous during the long-term culture and expansion of BM-MSCs [38, 39, 40], which display a fibroblast-like appearance at early passages while the flattened and larger morphology as well as a visible increase of cellular granularity in late passages. For example, one previous study showed that human BM-MSCs were consistent with a morphological appearance from passage 1 to passage 6–8 and beyond that period such cells became large and flat [40].

During further cultivation, MSCs demonstrate the altered common immunological surface markers. Comparison of the early and late passages of BM-MSCs reveals that no differences are observed between passage 2 and 6 MSCs in expression of CD44, CD90, CD105, HLA-ABC, and HLA-DR, while CD106 is downregulated in MSCs of passage 6 [41]. Research has also reported that the expression pattern of the common surface markers maintains consistently with consecutive passaging up to passage 8 of BM-MSCs [40]. In contrast, the positive expression of the common surface markers such as CD73, CD90 and CD105 presents at the passage 30 of human adipose-derived MSCs (AD-MSCs) [42] and human umbilical cord MSCs (UC-MSCs) [43].

4.2 Alterations of proliferation and differentiation of MSCs

MSCs exhibit a high proliferation rare at lower passage and, however, the rapid growth kinetics decrease gradually with consecutive cell passaging. A linear correlation is observed between cumulative population doubling and days in culture up to passage

6–8 of human BM-MSCs and the passage-dependent decrease in the proliferation rate is also observed beyond that period [40]. A reduction in the proliferation in the course of long-term cultivation has been reported in human dental pulp tissue-derived MSCs [44] and human tonsil-derived MSCs [45].

The differentiation ability of human BM-MSCs vary in long-term culture manifested by the significant reduction in expression levels of the osteogenic markers, such as ALP and osteocalcin, and adipogenic markers, such as fatty acid binding protein-4 and lipoprotein lipase at the late passages [45]. It has been reported that 25% samples of BM-MSCs from different donors in the 8th passage and the 20% in the 10th passage lost their osteogenic differentiation potential [46]. Similarly, 10% BM-MSC samples in the 6th passage, the 50% in the 8th passage, and the 60% in the 10th passage also lost their adipogenic differentiation [46]. In contrast, an in vitro differentiation study has also reported that the potential of adipogenesis decreases in higher passages (from the 5th passage) whereas the propensity for osteogenesis increases in the long-term culture [39].

4.3 Replicative senescence during long-term culture expansion of MSCs

Replicative senescence is the permanent cessation of growth in cells that divide, triggered by the shortening of telomeres. Senescence markers such senescence-associated β-galactosidase, heterochromatin protein-1, and p16INK4a are upregulated with age. Replicative senescence is the permanent cessation of growth in cells that divide, triggered by the shortening of telomeres. Senescence markers such senescence-associated β-galactosidase, heterochromatin protein-1, and p16INK4a are upregulated with age. [47, 48]. Molecular damage and epigenetic alterations occur in aging stem cells [49], which can result in the impairment of stem cell function.There are various signaling pathways involved in the senescence of MSCs, including oxidative damage [50, 51], age-related defects [52], and senescence associated up-regulation of microRNAs [53]. MSC senescence can be observed with long-term in vitro cultivation [54, 55], thus suggesting that a certain proportion of MSCs may undergo senescence during culture expansion. In vitro long-term culture of MSCs can induce continuous changes in gene expression [39, 56]. The expression levels of the senescence related genes, such as p16, p21 and p53, increase gradually in MSCs in the course of in vitro culture expansion [57]. DNA-methylation changes in MSCs during long-term culture have been investigated as an important epigenomic feature of replicative senescence of MSCs [58, 59, 60]. DNA-methylation changes may affect the proliferation and differentiation of MSCs. Differential methylation patterns of gene and miRNAs show between early-passage (passage 5) and late-passage (passage 15) MSCs [60]. Genes that are hypermethylated at passage 5 show reduced mRNA expression compared to those hypermethylated at passage 15, and vice versa. [60].

Genes that are hypermethylated at passage 5 show reduced mRNA expression compared to those hypermethylated at passage 15, and vice versa. [61, 62]. One previous report has shown that conditioned medium (CM) collected from senescent BM-MSC culture at passage 10 is able to trigger senescence in young cells [63]. The key factors of senescent MSC CM needed for triggering senescence in the young MSCs have been characterized as insulin-like growth factor binding proteins 4 and 7, which are linked to cellular senescence and apoptosis [63]. Similarly, monocyte chemoattractant proten-1 (MCP-1), as a dominant component of the SASP, is markedly increased in the conditioned medium of the late-phase MSCs and MCP-1 treatment significantly increase the senescence phenotypes of umbilical cord blood-derived MSCs via its cognate receptor chemokine receptor 2 signaling cascade [64].Metabolic changes are detected in the metabolome of MSCs as they undergo replicative senescence. This includes a decrease in nicotinamide ribonucleotide and an increase in orotic acid, which could serve as indicators of cellular senescence during the proliferation of MSCs in culture. [65].

4.4 In vitro long-term culture associated spontaneous transformation of MSCs

Sarcoma is a diverse collection of uncommon malignancies, and several research have explored the mesenchymal stem cell genesis of sarcoma. Various cellular and molecular mechanisms of mesenchymal stem cell (MSC) transformation are studied to gain a deeper insight into MSCs' role in sarcomagenesis. [66]. The majority of published research articles indicate that various sarcoma types have been shown MSCs' origin. Several group have reported spontaneous transformation in human and murine MSCs after long term culture [67, 68, 69, 70]. For example, one study has reported that murine BM-MSCs are spontaneously transformed at passage 29 under standard conditions and that these transformed MSCs are able to generate fibrosarcoma in immunocompromised mice [70]. Accumulated chromosomal abnormalities, such as chromosome instability, chromosomal imbalances and aneuploidy, are suggested to be associated with the transformation of BM MSCs [70]. Indeed, chromosomal aberrations (chromosomal level) in in vitro cultures of human MSCs have been reported in previous studies, including human BM-MSCs after passage 4 [71], human AD-MSCs from passage 5 [72], and UC-MSCs from passage 5 [73].There are also studies that have not detected the transformation of MSCs in long-term culture [74, 75, 76]. One previous study reports that human BM-MSCs do not undergo malignant transformation after long-term in vitro culture for up to 44 weeks and these cells maintain a normal karyotype [75]. In agree with the previous report [75], another study has described the occurrence of aneuploidy in cultivated human MSCs without evidence of transformation either in vitro or in vivo [76]. Røsland G.V. et al. found that human bone marrow-derived mesenchymal stem cells underwent spontaneous transformation due to cross-contamination with cancer cells [77]. Currently, there is no conclusive data supporting the conversion of various sarcoma subtypes from MSCs, which raises doubts about the potential of MSCs to change spontaneously.

5. CONCISE SUMMARY

MSCs have significant potential in translational medicine for addressing many diseases or medical problems. Mesenchymal stem cells (MSCs) are multipotent stem cells that can be extracted from many organs, with bone marrow being a significant reservoir of human MSCs. Since a substantial quantity of MSCs is needed for therapeutic use, it is crucial to expand MSCs in vitro. MSCs at greater passage may result in culture-induced changes, including variations in cellular shape, immunological surface markers, proliferation, differentiation, and cell genetics. The safety of MSC therapy is a serious concern due to the spontaneous change of MSCs during long-term in vitro cultivation. Human mesenchymal stem cells (MSCs) from different organs have diverse biological characteristics. Currently, there is no agreement on the materials and culture techniques, culture conditions, growth factor supplements, freezing and thawing methods (e.g. medium), and functional tests. Standardizing the processes for preparing MSCs is necessary for pre-clinical investigations and clinical use of MSCs.

6. AT A GLANCE

Mesenchymal stem cell/stromal cells (MSCs) have the ability to transform into numerous cell types such as osteocytes, adipocytes, and chondrocytes. MSCs are found in several adult tissues including bone marrow, adipose tissue, and different neonatal birth-related tissues. Due to their ability to self-renew and differentiate, as well as their immunomodulatory and paracrine capabilities, and the absence of major histocompatibility complex (MHC) class II molecules, MSCs have garnered significant interest in stem cell-based translational medicine research. Because of their rarity in various tissues, MSCs can be extracted and cultured in a laboratory setting to generate an ample number of cells for use in clinical settings. This chapter will outline the procedure for obtaining primary bone marrow-derived MSCs and their in vitro culture growth. Functional qualities are evaluated by differentiation assays such as osteogenesis, chondrogenesis, and adipogenesis. Additionally, 3-D culture of MSCs and co-culture of MSCs with tumor cells are available. This discussion will focus on the long-term changes in MSCs, such as replicative senescence and spontaneous transformation, to enhance knowledge of using MSCs in the initial phases for safe and efficient cell-based therapy.

ABBREVIATIONS

AD-MSCs	Adipose-derived mesenchymal stem cell/stromal cells
ALP	Alkaline phosphatase
AsAP	Ascorbic acid phosphate
BM-MSCs	Bone marrow-derived MSCs
CM	Conditioned medium
CFU-F	Colony forming unit-fibroblast
DMEM	Dulbecco's modified Eagle's medium
DMSO	Dimethyl sulphoxide
EDTA	Ethylenediaminetetraacetic acid
FBS	Fetal bovine serum
ISCT	International Society for Cellular Therapy
MCP-1	Monocyte chemoattractant proten-1
MHC	Major histocompatibility complex
MSCs	Mesenchymal stem cell/stromal cells
PBS	Phosphate buffered saline
SASP	Senescence-associated secretory phenotype
UC-MSCs	Umbilical cord mesenchymal stem cell/stromal cells

REFERENCES

1. Friedenstein AJ, Piatetzky-Shapiro II, Petrakova KV. Osteogenesis in transplants of bone marrow cells. J Embryol Exp Morphol. 1966;16:381-390

2. Friedenstein AJ, Chailakhjan RK, Lalykina KS. The development of fibroblast colonies in monolayer cultures of guinea-pig bone marrow and spleen cells. Cell Tissue Kinet. 1970;3:393-403. DOI: 10.1111/j.1365-2184.1970.tb00347.x

3. Prockop DJ. Marrow stromal cells as stem cells for nonhematopoietic tissues. Science. 1997;276(5309):71-74. DOI: 10.1126/science.276.5309.71

4. Caplan AI. Mesenchymal stem cells. J Orthop Res. 1991;9:641-650. doi: 10.1002/jor.1100090504

5. Dominici M, Le Blanc K, Mueller I, Slaper-Cortenbach I, Marini F, Krause D, Deans R, Keating A, Prockop Dj, Horwitz E. Minimal criteria for defining multipotent mesenchymal stromal cells. The International Society for Cellular Therapy position statement. Cytotherapy. 2006;8:315-317. doi: 10.1080/14653240600855905

6. Kozlowska U, Krawczenko A, Futoma K, Jurek T, Rorat M, Patrzalek D, Klimczak A. Similarities and differences between mesenchymal stem/progenitor cells derived from various human tissues. World J Stem Cells. 2019;11:347-374. doi: 10.4252/wjsc.v11.i6.347

7. Melief SM, Zwaginga JJ, Fibbe WE, Roelofs H. Adipose tissue-derived multipotent stromal cells have a higher immunomodulatory capacity than their bone marrow-derived counterparts. Stem Cells Transl Med. 2013;2:455-463. doi: 10.5966/sctm.2012-0184

8. Phinney DG, Sensebé L. Mesenchymal stromal cells: misconceptions and evolving concepts. Cytotherapy. 2013;15:140-145. doi: 10.1016/j.jcyt.2012.11.005

9. Viswanathan S, Shi Y, Galipeau J, Krampera M, Leblanc K, Martin I, Nolta J, Phinney DG, Sensebe L. Mesenchymal stem versus stromal cells: International Society for Cell & Gene Therapy (ISCT®) Mesenchymal Stromal Cell committee position statement on nomenclature. Cytotherapy. 2019;21:1019-1024. doi: 10.1016/j.jcyt.2019.08.002

10. Du WJ, Chi Y, Yang ZX, Li ZJ, Cui JJ, Song BQ, Li X, Yang SG, Han ZB, Han ZC. Heterogeneity of proangiogenic features in mesenchymal stem cells derived from bone marrow, adipose

tissue, umbilical cord, and placenta. Stem Cell Res Ther. 2016;7:163. doi: 10.1186/s13287-016-0418-9

11. Ricciardi M, Malpeli G, Bifari F, Bassi G, Pacelli L, Nwabo Kamdje AH, Chilosi M, Krampera M. Comparison of epithelial differentiation and immune regulatory properties of mesenchymal stromal cells derived from human lung and bone marrow. PLoS One. 2012;7:e35639. doi: 10.1371/journal.pone.0035639

12. Drela K, Stanaszek L, Snioch K, Kuczynska Z, Wrobel M, Sarzynska S, Legosz P, Maldyk P, Lukomska B. Bone marrow-derived from the human femoral shaft as a new source of mesenchymal stem/stromal cells: an alternative cell material for banking and clinical transplantation. Stem Cell Res Ther. 2020;11:262. doi: 10.1186/s13287-020-01697-5

13. Priya N, Sarcar S, Majumdar AS, SundarRaj S. Explant culture: a simple, reproducible, efficient and economic technique for isolation of mesenchymal stromal cells from human adipose tissue and lipoaspirate. J Tissue Eng Regen Med. 2014;8:706-716. doi: 10.1002/term.1569

14. Nepali S, Park M, Lew H, Kim O. Comparative Analysis of Human Adipose-Derived Mesenchymal Stem Cells from Orbital and Abdominal Fat. Stem Cells Int. 2018;2018:3932615. doi: 10.1155/2018/3932615

15. Steens J, Klar L, Hansel C, Slama A, Hager T, Jendrossek V, Aigner C, Klein D. The vascular nature of lung-resident mesenchymal stem cells. Stem Cells Transl Med. 2021;10:128-143. doi: 10.1002/sctm.20-0191

16. Ferro T, Santhagunam A, Madeira C, Salgueiro JB, da Silva CL, Cabral JMS. Successful isolation and ex vivo expansion of human mesenchymal stem/stromal cells obtained from different synovial tissue-derived (biopsy) samples. J Cell Physiol. 2019;234:3973-3984. doi: 10.1002/jcp.27202

17. De Bari C, Dell'Accio F, Tylzanowski P, Luyten FP. Multipotent mesenchymal stem cells from adult human synovial membrane. Arthritis Rheum. 2001;44:1928-1942. doi: 10.1002/1529-0131(200108)44:8<1928::AID-ART331>3.0.CO;2-P

18. Lei M, Li K, Li B, Gao LN, Chen FM, Jin Y. Mesenchymal stem cell characteristics of dental pulp and periodontal ligament stem cells after in vivo transplantation. Biomaterials. 2014;35:6332-6343. doi: 10.1016/j.biomaterials.2014.04.071

19. Abdulrazzak H, Moschidou D, Jones G, Guillot PV. Biological characteristics of stem cells from foetal, cord blood and extraembryonic tissues. J R Soc Interface. 2010;7 Suppl 6(Suppl 6):S689-706. doi: 10.1098/rsif.2010.0347.focus

20. Cronwright G, Le Blanc K, Götherström C, Darcy P, Ehnman M, Brodin B. Cancer/testis antigen expression in human mesenchymal stem cells: down-regulation of SSX impairs cell migration and matrix metalloproteinase 2 expression. Cancer Res. 2005;65:2207-2215. doi: 10.1158/0008-5472.CAN-04-1882

21. Wang XY, Lan Y, He WY, Zhang L, Yao HY, Hou CM, Tong Y, Liu YL, Yang G, Liu XD, Yang X, Liu B, Mao N. Identification of mesenchymal stem cells in aorta-gonad-mesonephros and yolk sac of human embryos. Blood. 2008;111:2436-2443. doi: 10.1182/blood-2007-07-099333

22. Phunikom N, Boonmuen N, Kheolamai P, Suksen K, Manochantr S, Tantrawatpan C, Tantikanlayaporn D. Andrographolide promotes proliferative and osteogenic potentials of human placenta-derived mesenchymal stem cells through the activation of Wnt/β-catenin signaling. Stem Cell Res Ther. 2021;12:241. doi: 10.1186/s13287-021-02312-x

23. Araújo AB, Salton GD, Furlan JM, Schneider N, Angeli MH, Laureano ÁM, Silla L, Passos EP, Paz AH. Comparison of human mesenchymal stromal cells from four neonatal tissues: Amniotic membrane, chorionic membrane, placental decidua and

umbilical cord. Cytotherapy. 2017;19:577-585. doi: 10.1016/j.jcyt.2017.03.001

24. Ma J, Wu J, Han L, Jiang X, Yan L, Hao J, Wang H. Comparative analysis of mesenchymal stem cells derived from amniotic membrane, umbilical cord, and chorionic plate under serum-free condition. Stem Cell Res Ther. 2019;10:19. doi: 10.1186/s13287-018-1104-x

25. Semenova E, Grudniak MP, Machaj EK, Bocian K, Chroscinska-Krawczyk M, Trochonowicz M, Stepaniec IM, Murzyn M, Zagorska KE, Boruczkowski D, Kolanowski TJ, Oldak T, Rozwadowska N. Mesenchymal Stromal Cells from Different Parts of Umbilical Cord: Approach to Comparison & Characteristics. Stem Cell Rev Rep. doi: 10.1007/s12015-021-10157-3

26. Zhao D, Liu L, Chen Q, Wang F, Li Q, Zeng Q, Huang J, Luo M, Li W, Zheng Y, Liu T. Hypoxia with Wharton's jelly mesenchymal stem cell coculture maintains stemness of umbilical cord blood-derived CD34+ cells. Stem Cell Res Ther. 2018;9:158. doi: 10.1186/s13287-018-0902-5

27. Amati E, Perbellini O, Rotta G, Bernardi M, Chieregato K, Sella S, Rodeghiero F, Ruggeri M, Astori G. High-throughput immunophenotypic characterization of bone marrow- and cord blood-derived mesenchymal stromal cells reveals common and differentially expressed markers: identification of angiotensin-converting enzyme (CD143) as a marker differentially expressed between adult and perinatal tissue sources. Stem Cell Res Ther. 2018;16;9:10. doi: 10.1186/s13287-017-0755-3

28. Ilmer M, Karow M, Geissler C, Jochum M, Neth P. Human osteoblast-derived factors induce early osteogenic markers in human mesenchymal stem cells. Tissue Eng Part A. 2009;15:2397-2409. doi: 10.1089/ten.tea.2008.0427

29. Li C, Sunderic K, Nicoll SB, Wang S. Downregulation of Heat Shock Protein 70 Impairs Osteogenic and Chondrogenic

Differentiation in Human Mesenchymal Stem Cells. Sci Rep. 2018;8:553. doi: 10.1038/s41598-017-18541-1

30. Chen J, Shi ZD, Ji X, Morales J, Zhang J, Kaur N, Wang S. Enhanced osteogenesis of human mesenchymal stem cells by periodic heat shock in self-assembling peptide hydrogel. Tissue Eng Part A. 2013;19:716-728. doi: 10.1089/ten.TEA.2012.0070

31. Moradi F, Bahktiari M, Joghataei MT, Nobakht M, Soleimani M, Hasanzadeh G, Fallah A, Zarbakhsh S, Hejazian LB, Shirmohammadi M, Maleki F. BD PuraMatrix peptide hydrogel as a culture system for human fetal Schwann cells in spinal cord regeneration. J Neurosci Res. 2012;90:2335-2348. doi: 10.1002/jnr.23120

32. Chen J, Li C, Wang S. Periodic heat shock accelerated the chondrogenic differentiation of human mesenchymal stem cells in pellet culture. PLoS One. 2014;9:e91561. doi: 10.1371/journal.pone.0091561

33. Luetzkendorf J, Nerger K, Hering J, Moegel A, Hoffmann K, Hoefers C, Mueller-Tidow C, Mueller LP. Cryopreservation does not alter main characteristics of Good Manufacturing Process-grade human multipotent mesenchymal stromal cells including immunomodulating potential and lack of malignant transformation. Cytotherapy. 2015;17:186-198. doi: 10.1016/j.jcyt.2014.10.018

34. Waterman RS, Tomchuck SL, Henkle SL, Betancourt AM. A new mesenchymal stem cell (MSC) paradigm: polarization into a pro-inflammatory MSC1 or an Immunosuppressive MSC2 phenotype. PLoS One. 2010;5:e10088. doi: 10.1371/journal.pone.0010088

35. Marinescu CI, Preda MB, Burlacu A. A procedure for in vitro evaluation of the immunosuppressive effect of mouse mesenchymal stem cells on activated T cell proliferation. Stem Cell Res Ther. 2021;12:319. doi: 10.1186/s13287-021-02344-3

36. Jing D, Fonseca AV, Alakel N, Fierro FA, Muller K, Bornhauser M, Ehninger G, Corbeil D, Ordemann R. Hematopoietic stem cells in co-culture with mesenchymal stromal cells--modeling the niche compartments in vitro. Haematologica. 2010;95:542-550. doi: 10.3324/haematol.2009.010736

37. Joshi A, Xu Z, Ikegami Y, Yamane S, Tsurashima M, Ijima H. Co-culture of mesenchymal stem cells and human umbilical vein endothelial cells on heparinized polycaprolactone/gelatin co-spun nanofibers for improved endothelium remodeling. Int J Biol Macromol. 2020;151:186-192. doi: 10.1016/j.ijbiomac.2020.02.163

38. Redaelli S, Bentivegna A, Foudah D, Miloso M, Redondo J, Riva G, Baronchelli S, Dalprà L, Tredici G. From cytogenomic to epigenomic profiles: monitoring the biologic behavior of in vitro cultured human bone marrow mesenchymal stem cells. Stem Cell Res Ther. 2012;3:47. doi: 10.1186/scrt138

39. Wagner W, Horn P, Castoldi M, Diehlmann A, Bork S, Saffrich R, Benes V, Blake J, Pfister S, Eckstein V, Ho AD. Replicative senescence of mesenchymal stem cells: a continuous and organized process. PLoS One. 2008;3:e2213. doi: 10.1371/journal.pone.0002213

40. Khoo ML, Shen B, Tao H, Ma DD. Long-term serial passage and neuronal differentiation capability of human bone marrow mesenchymal stem cells. Stem Cells Dev. 2008;17:883-896. doi: 10.1089/scd.2007.0185

41. Halfon S, Abramov N, Grinblat B, Ginis I. Markers distinguishing mesenchymal stem cells from fibroblasts are downregulated with passaging. Stem Cells Dev. 2011;20:53-66. doi: 10.1089/scd.2010.0040

42. Danisovic L, Oravcova L, Krajciova L, Varchulova Novakova Z, Bohac M, Varga I, Vojtassak J. Effect of long-term culture on the biological and morphological characteristics of human adipose tissue-derived stem Cells. J Physiol Pharmacol. 2017;68:149-158

43. Wang Y, Zhang Z, Chi Y, Zhang Q, Xu F, Yang Z, Meng L, Yang S, Yan S, Mao A, Zhang J, Yang Y, Wang S, Cui J, Liang L, Ji Y, Han ZB, Fang X, Han ZC. Long-term cultured mesenchymal stem cells frequently develop genomic mutations but do not undergo malignant transformation. Cell Death Dis. 2013;4:e950. doi: 10.1038/cddis.2013.480

44. Wang H, Zhong Q, Yang T, Qi Y, Fu M, Yang X, Qiao L, Ling Q, Liu S, Zhao Y. Comparative characterization of SHED and DPSCs during extended cultivation in vitro. Mol Med Rep. 2018;17:6551-6559. doi: 10.3892/mmr.2018.8725

45. Yu Y, Park YS, Kim HS, Kim HY, Jin YM, Jung SC, Ryu KH, Jo I. Characterization of long-term in vitro culture-related alterations of human tonsil-derived mesenchymal stem cells: role for CCN1 in replicative senescence-associated increase in osteogenic differentiation. J Anat. 2014;225:510-518. doi: 10.1111/joa.12229

46. Gharibi B, Hughes FJ. Effects of medium supplements on proliferation, differentiation potential, and in vitro expansion of mesenchymal stem cells. Stem Cells Transl Med. 2012;1:771-782. doi: 10.5966/sctm.2010-0031

47. Sharpless NE, DePinho RA. How stem cells age and why this makes us grow old. Nat Rev Mol Cell Biol. 2007;8:703-713. doi: 10.1038/nrm2241

48. Liu L, Rando TA. Manifestations and mechanisms of stem cell aging. J Cell Biol. 2011;193: 257-266. doi: 10.1083/jcb.201010131

49. Ermolaeva M, Neri F, Ori A, Rudolph KL. Cellular and epigenetic drivers of stem cell ageing. Nat Rev Mol Cell Biol. 2018;19:594-610. doi: 10.1038/s41580-018-0020-3

50. Alves H, Munoz-Najar U, De Wit J, Renard AJ, Hoeijmakers JH, Sedivy JM, Van Blitterswijk C, De Boer J. A link between the accumulation of DNA damage and loss of multi-potency of human mesenchymal stromal cells. J Cell Mol Med. 2010;14:2729-2738. doi: 10.1111/j.1582-4934.2009.00931.x

51. Ko E, Lee KY, Hwang DS. Human umbilical cord blood-derived mesenchymal stem cells undergo cellular senescence in response to oxidative stress. Stem Cells Dev. 2012;21:1877-1886. doi: 10.1089/scd.2011.0284

52. Stolzing A, Jones E, McGonagle D, Scutt A. Age-related changes in human bone marrow-derived mesenchymal stem cells: consequences for cell therapies. Mech Ageing Dev. 2008;129:163-173. doi: 10.1016/j.mad.2007.12.002

53. Cakouros D, Gronthos S. The changing epigenetic landscape of Mesenchymal Stem/Stromal Cells during aging. Bone. 2020;137:115440. doi: 10.1016/j.bone.2020.115440

54. Bonab MM, Alimoghaddam K, Talebian F, Ghaffari SH, Ghavamzadeh A, Nikbin B. Aging of mesenchymal stem cell in vitro. BMC Cell Biol. 2006;7:14. doi: 10.1186/1471-2121-7-14

55. Turinetto V, Vitale E, Giachino C. Senescence in human mesenchymal stem cells: Functional changes and implications in stem cell-based therapy. Int J Mol Sci. 2016;17:1164. doi: 10.3390/ijms17071164

56. Wagner W, Bork S, Lepperdinger G, Joussen S, Ma N, Strunk D, Koch C. How to track cellular aging of mesenchymal stromal cells? Aging (Albany NY). 2010;2:224-230. doi: 10.18632/aging.100136

57. Cheng H, Qiu L, Ma J, Zhang H, Cheng M, Li W, Zhao X, Liu K. Replicative senescence of human bone marrow and umbilical cord derived mesenchymal stem cells and their differentiation to adipocytes and osteoblasts. Mol Biol Rep. 2011;38:5161-5168. doi: 10.1007/s11033-010-0665-2

58. Bork S, Pfister S, Witt H, Horn P, Korn B, Ho AD, Wagner W. DNA methylation pattern changes upon long-term culture and aging of human mesenchymal stromal cells. Aging Cell. 2010;9:54-63. doi: 10.1111/j.1474-9726.2009.00535.x

59. Schellenberg A, Lin Q, Schüler H, Koch CM, Joussen S, Denecke B, Walenda G, Pallua N, Suschek CV, Zenke M, Wagner W. Replicative senescence of mesenchymal stem cells causes DNA-methylation changes which correlate with repressive histone marks. Aging (Albany NY). 2011;3:873-888. doi: 10.18632/aging.100391

60. Choi MR, In YH, Park J, Park T, Jung KH, Chai JC, Chung MK, Lee YS, Chai YG. Genome-scale DNA methylation pattern profiling of human bone marrow mesenchymal stem cells in long-term culture. Exp Mol Med. 2012;44:503-512. doi: 10.3858/emm.2012.44.8.057

61. Herranz N, Gil J. Mechanisms and functions of cellular senescence. J Clin Invest. 2018;128:1238-1246. doi: 10.1172/JCI95148

62. Di Micco R, Krizhanovsky V, Baker D, d'Adda di Fagagna F. Cellular senescence in ageing: from mechanisms to therapeutic opportunities. Nat Rev Mol Cell Biol. 2021;22:75-95. doi: 10.1038/s41580-020-00314-w

63. Severino V, Alessio N, Farina A, Sandomenico A, Cipollaro M, Peluso G, Galderisi U, Chambery A. Insulin-like growth factor binding proteins 4 and 7 released by senescent cells promote premature senescence in mesenchymal stem cells. Cell Death Dis. 2013;4:e911. doi: 10.1038/cddis.2013.445

64. Jin HJ, Lee HJ, Heo J, Lim J, Kim M, Kim MK, Nam HY, Hong GH, Cho YS, Choi SJ, Kim IG, Shin DM, Kim SW. Senescence-associated MCP-1 secretion is dependent on a decline in BMI1 in human mesenchymal stromal cells. Antioxid Redox Signal. 2016;24:471-485. doi: 10.1089/ars.2015.6359

65. Fernandez-Rebollo E, Franzen J, Goetzke R, Hollmann J, Ostrowska A, Oliverio M, Sieben T, Rath B, Kornfeld JW, Wagner W. Senescence-associated metabolomic phenotype in primary and iPSC-derived mesenchymal stromal cells. Stem Cell Reports. 2020; 14: 201-209. doi: 10.1016/j.stemcr.2019.12.012

66. Li C, Zhao H, Wang B. Mesenchymal stem/stromal cells: Developmental origin, tumorigenesis and translational cancer therapeutics. Transl Oncol. 2021;14:100948. doi: 10.1016/j.tranon.2020.100948

67. Tang Q, Chen Q, Lai X, Liu S, Chen Y, Zheng Z, Xie Q, Maldonado M, Cai Z, Qin S, Ho G, Ma L. Malignant transformation potentials of human umbilical cord mesenchymal stem cells both spontaneously and via 3-methycholanthrene induction. PLoS One. 2013;8:e81844. doi: 10.1371/journal.pone.0081844

68. Pan Q, Fouraschen SM, de Ruiter PE, Dinjens WN, Kwekkeboom J, Tilanus HW, van der Laan LJ. Detection of spontaneous tumorigenic transformation during culture expansion of human mesenchymal stromal cells. Exp Biol Med (Maywood). 2014;239:105-115. doi: 10.1177/1535370213506802

69. Li H, Fan X, Kovi RC, Jo Y, Moquin B, Konz R, Stoicov C, Kurt-Jones E, Grossman SR, Lyle S, Rogers AB, Montrose M, Houghton J. Spontaneous expression of embryonic factors and p53 point mutations in aged mesenchymal stem cells: a model of age-related tumorigenesis in mice. Cancer Res. 2007;67:10889-10898. doi: 10.1158/0008-5472.CAN-07-2665

70. Miura M, Miura Y, Padilla-Nash HM, Molinolo AA, Fu B, Patel V, Seo BM, Sonoyama W, Zheng JJ, Baker CC, Chen W, Ried T, Shi S. Accumulated chromosomal instability in murine bone marrow mesenchymal stem cells leads to malignant transformation. Stem Cells. 2006;24:1095-1103. doi: 10.1634/stemcells.2005-0403

71. Binato R, de Souza Fernandez T, Lazzarotto-Silva C, Du Rocher B, Mencalha A, Pizzatti L, Bouzas LF, Abdelhay E. Stability of human mesenchymal stem cells during in vitro culture: considerations for cell therapy. Cell Prolif. 2013;46:10-22. doi: 10.1111/cpr.12002

72. Froelich K, Mickler J, Steusloff G, Technau A, Ramos Tirado M, Scherzed A, Hackenberg S, Radeloff A, Hagen R, Kleinsasser N. Chromosomal aberrations and deoxyribonucleic acid single-strand breaks in adipose-derived stem cells during long-term expansion in vitro. Cytotherapy. 2013;15:767-781. doi: 10.1016/j.jcyt.2012.12.009

73. Borghesi A, Avanzini MA, Novara F, Mantelli M, Lenta E, Achille V, Cerbo RM, Tzialla C, Longo S, De Silvestri A, Zimmermann LJ, Manzoni P, Zecca M, Spinillo A, Maccario R, Zuffardi O, Stronati M. Genomic alterations in human umbilical cord-derived mesenchymal stromal cells call for stringent quality control before any possible therapeutic approach. Cytotherapy. 2013;15:1362-1373. doi: 10.1016/j.jcyt.2013.06.006

74. Chen G, Yue A, Ruan Z, Yin Y, Wang R, Ren Y, Zhu L. Human umbilical cord-derived mesenchymal stem cells do not undergo malignant transformation during long-term culturing in serum-free medium. PLoS One. 2014;9:e98565. doi: 10.1371/journal.pone.0098565

75. Bernardo ME, Zaffaroni N, Novara F, Cometa AM, Avanzini MA, Moretta A, Montagna D, Maccario R, Villa R, Daidone MG, Zuffardi O, Locatelli F. Human bone marrow derived mesenchymal stem cells do not undergo transformation after long-term in vitro culture and do not exhibit telomere maintenance mechanisms. Cancer Res. 2007;67:9142-9149. doi: 10.1158/0008-5472.CAN-06-4690

76. Tarte K, Gaillard J, Lataillade JJ, Fouillard L, Becker M, Mossafa H, Tchirkov A, Rouard H, Henry C, Splingard M, Dulong J, Monnier D, Gourmelon P, Gorin NC, Sensebé L; Société Française de Greffe de Moelle et Thérapie Cellulaire. Clinical-grade production of human mesenchymal stromal cells: occurrence of aneuploidy without transformation. Blood. 2010;115:1549-1553. doi: 10.1182/blood-2009-05-219907

77. Torsvik A, Røsland GV, Svendsen A, Molven A, Immervoll H, McCormack E, Lønning PE, Primon M, Sobala E, Tonn JC, Goldbrunner R, Schichor C, Mysliwietz J, Lah TT, Motaln H, Knappskog S, Bjerkvig R. Spontaneous malignant transformation of human mesenchymal stem cells reflects cross-contamination: putting the research field on track - letter. Cancer Res. 2010;70:6393-6396. doi: 10.1158/0008-5472. CAN-10-1305

CHAPTER 5

UTILIZING STABLE ISOTOPE LABELING WITH AMINO ACIDS IN CELL CULTURE (SILAC) TO INVESTIGATE IVERMECTIN INDUCED MOLECULAR PATHWAY ALTERATIONS IN HUMAN OVARIAN CANCER CELLS

1. INTRODUCTION: AN OVERVIEW

SILAC is a polypeptide-labeling method created by the Thermo Fisher company in the United States in 2002[1]. Heavy isotopes (13C or 15N) and light isotopes (12C or 14N) identify two important amino acids (L-lysine and L-arginine) in a cell-cultured media. Following 6-10 generations of cell culture with necessary amino acids, all proteins were tagged with heavy or light isotopes. Mass spectrometry (MS) is used to study the cellular proteins affected by various treatment parameters to gather qualitative and quantitative proteome data[2]. SILAC generally allows heavy and light isotopes-labeled sample cells at the early stage of the experimental workflow, so the variability caused by the sample handling process was minimized [3]. SILAC was widely used in quantitative proteomics to study pathogenesis, drug target, protein modification and dynamics, protein-molecule interaction, and screen special functional proteins [4]. SILAC showed outstanding performance for quantification and dynamics of phosphosites in colorectal cancer with the treatment of the epidermal growth factor receptor (EGFR)-blocking antibody cetuximab, rendering it the effective method for cellular signaling study in cell culture models [5]. In terms of identification of protein-molecule interaction, SILAC combined with various affinity purifications of protein experimental setups could be used to distinguish specific complexes from nonspecific ones [6]. One study performed SILAC to overcome the most challenging problem in defining specific partners in protein complexes. The cells containing an affinity tagged protein were cultured in a light isotopic medium, while wild-type cells were grown in a heavy isotopic medium. The results of MS showed that specific partners appear as isotopically light [7]. SILAC also offers numerous opportunities to discover potential biomarkers and therapeutic targets for some drugs [8]. SILAC in combination with other developed approaches made SILAC more popular; for example, these SILAC labels in pulse or pulse-chase scenarios could be used to measure macromolecular dynamics on time scales of several hours [9]. A method using dynamic-SILAC labeling combined with isobaric mass tagging was effectively utilized to study protein breakdown and production in cellular systems [10]. SILAC offered an efficient method to thoroughly and systematically assess and measure the intricate proteome of mammalian cells, which could advance developments in the medical industry. Ivermectin, marketed in 1981, was commonly used as a broad-spectrum antiparasitic compound. It was approved to treat onchocerciasis (150–200 μg kg−1 body weight), scabies (200 μg kg−1 body weight), lymphatic filariasis (150–200 μg kg−1 body weight), demodicosis (200 μg kg−1 body weight), strongyloidiasis (200 μg kg−1 body weight), pediculosis (400 μg kg−1 body weight), and filariasis (due to Mansonella ozzardi, 6 mg as a single dose) [11].

Due to its primary action on chloride-dependent channels such as γ-aminobutyric acid and glutamate, ivermectin's safety profile in larger animals may be acceptable. The blood-brain barrier in humans can limit the transport of ivermectin to the central nervous system. [12]. The safety of ivermectin has been proved with clinical studies on children, infants, and pregnant women. A study including 170 infants and children with the treatment of oral ivermectin (mean dose = 223 μg kg−1) showed good tolerance, and only seven subjects occurred mild adverse events [13]. A study including 893 pregnant women with the oral treatment of ivermectin also showed good tolerance, and no patients were reported to generate serious events (stillbirths, neonatal death, low birth weight, spontaneous abortions, preterm births, and congenital anomalies) [14]. Those studies proved that ivermectin was safe enough to be used in human diseases, but there was still insufficient evidence to prove no adverse side effects. The highest ivermectin dose was 200 μg/kg, which was approved by FDA. However, some patients without serious events have used 10 times more than the FDA-approved dose [15]. All those made ivermectin more likely to achieve success in clinical application. In recent days, studies found that ivermectin was effective in a completely new range of diseases, such as neurological disorders, antiviral (e.g., dengue, HIV, and encephalitis), antibacterial (e.g., Buruli ulcer and tuberculosis), anticancer (melanoma, lymphoid leukemia, lung cancer, glioblastoma, and breast cancer) [11]. The functions and mechanisms of ivermectin on anticancer generated interest and excitement in the scientific community. Ivermectin suppresses breast cancer by disrupting cellular signaling in the process and activating cytostatic autophagy through mediating PAK1 expression [16]. Ivermectin showed a synergistic effect with the chemotherapy agents by increasing cell death in leukemia cells. Some researchers, who aimed at overcoming cancer, claimed that ivermectin could be rapidly advanced into clinical trials [17]. Further study on molecular network, signaling pathway, and key biological processes of ivermectin would provide more useful information about this multifaceted "wonder" drug. This chapter explains how SILAC is used to identify differentially expressed proteins (DEGs) in ovarian cancer cells treated with ivermectin. 1. Cultured ovarian cancer cells (TOV21G) were labeled with heavy and light SILAC reagents. 2. The SILAC-labeled TOV21G cells were treated with ivermectin and proteins were prepared. 3. The quality of SILAC-labeled protein samples was assessed using 1D SDS-PAGE. 4. SILAC-labeled proteins were digested with trypsin. 5. Each fraction underwent LC-MS/MS analysis. 6. Bioinformatics analysis was conducted on signaling pathways and biological processes. SILAC is a valuable tool for identifying protein modifications and dynamic alterations in living cells. The findings can offer scientific insights into the molecular mechanisms of ivermectin in ovarian cancer.

2. METHODS

2.1 Ovarian cancer cell culture

The human ovarian cell line (TOV-21G) was cultured in RPMI 1640 medium supplemented with 10% fetal bovine serum (FBS) and 5% CO2 at 37°C. (i) Ovarian cell line used here was TOV-21G, which was obtained from Keibai Academy of Science (Nanjing, China) [18]. (ii) RPMI 1640 was used without glutamine, lysine, and arginine. (iii) FBS was brought from Gibco® Certified Thermo Fisher Scientific. (iv) The growing states of TOV-21G were observed, and the medium was changed in every 2 days.

2.2 SILAC medium preparation and labeling

SILAC "light" or "heavy" labeling growing medium (Thermo Fisher Scientific, US) was used to culture TOV-21G cells for 10 passages, to ensure a high level of stable isotope replacing original amino acids [8]. (i) "Light" labeled amino acids: 50 mg L-arginine HCl (Arg0), 50 mg L-lysine HCl (Lys0). "Heavy" labeled amino acids: 50 mg L-arginine-13C6,15N4 HCl (Arg10), 50 mg L-lysine-13C6,15N2 HCl (Lys8). (ii) A total of 1 L RPMI 1640 without glutamine, lysine, and arginine. (iii) For SILAC experiments, 50 mg Arg0 and 50 mg

Lys0 ("light" labeling reagent) were added into 500 mL RPMI 1640 medium to form SILAC "light" labeling growing medium. A total of 50 mg Arg10 and 50 mg Lys8 ("heavy" labeling reagent) were added to 500 mL RPMI 1640 medium to form SILAC "heavy" labeling growing medium. (iv) The growing states of TOV-21G were observed, and the medium was changed in every 2 days. TOV-21G cells were cultured and passaged for 10 generations in 10-cm culture flasks.

2.3 Ivermectin treatment of SILAC-labeled TOV21G cells

When SILAC-labeled TOV21G cells achieved 80% cell density in 10-cm culture flasks, 20 μM ivermectin was added to TOV-21G cells in SILAC "heavy" labeling growing medium, and 0.1% DMSO was added to TOV-21G cells in SILAC "light" labeling growing medium for 24 h [19]. (i) Ivermectin ($C_{48}H_{74}O_{14}$, purity ≥ 95%): The drug was brought from Solarbio (http://www.solarbio.com/goods-3911.html). (ii) TOV-21G cells were counted, and 8000 cells/well were seeded in 96-well plates. Ivermectin was added into each well in different drug concentrations (0 μM, 10 μM, 20 μM, 30 μM, 40 μM, and 50 μM) for 24 h. CCK8 (10 μL) was added into each well for 1 h to measure absorbance values (OD) at a wavelength of 450 nm. Lethal concentration 50 (IC50) was calculated according to OD values of each well in different ivermectin concentrations. (iii) Ivermectin treatment group: 20 μM ivermectin was added to TOV-21G cells in SILAC "heavy" labeling growing medium for 24 h. Control group: 0.1% DMSO was added to TOV-21G cells in SILAC "light" labeling growing medium for 24 h.

2.4 Cell lysis and protein preparation

Ivermectin treatment and 0.1% DMSO treatment TOV-21G cells were collected and lysed by protein isolation buffer, respectively. (i) TOV-21G cells collection: A total of 500 μL trypsin was added to each 10-cm culture flasks for several minutes and collected with centrifugation (800 × g, 5 min). TOV-21G cells were washed with ice-cold phosphate buffer solution (PBS) for three times. (ii) Protein isolation buffer: 2 mM thiourea, 4% CHAPS (3-[(3-cholamidopropyl)-

dimethylammonio] -1-propane), 7 M urea, 100 mM dithiothreitol (DTT), and 2% ampholyte. (iii) TOV-21G cells lysis: A total of 200 μL protein isolation buffer was added to each 10-cm culture flasks for 30 min (ice-cold) and then oscillated with five vortex cycles. (iv) SILAC-labeled protein collection: Protein isolation buffer was centrifuged (13,000 × g, 20 min, 4°C), and the SILAC-labeled protein samples were collected from the supernatants in new tubes. (v) Protein concentrations measurement: Protein concentrations of the SILAC-labeled protein samples were measured with the 2-D quant protein assay kit (Bio-Rad, US).

2.5 The preliminary experiment of SILAC-labeled protein samples

The "heavy"- and "light"-SILAC-labeled proteins were mixed and loaded onto 1X SDS-PAGE to check the quality. SDS-PAGE-separated proteins were further analyzed with MS/MS as a preliminary experiment to check the labeling efficiency. (i) The loading sample preparation: According to the 1:1 ratio, the "heavy"- and "light"-SILAC-labeled proteins were mixed in a 5X loading buffer. (ii) Electrophoresis: The mixed SILAC-labeled proteins were loaded onto SDS-PAGE gel (gel concentration: 12.5%) with the amount of 20 μg/lane by constant current (14 mA, 90 min). (iii) Coomassie brilliant blue staining: Prepare Coomassie brilliant blue stain and destain solutions. Filter the stain solution through Whatman 1 filter paper. (iv) MS/MS: Proteins were separated from SDS-PAGE bands, and then were reduced, alkylated, and trypsin-digested. The tryptic peptides were analyzed with MS/MS.

2.6 Trypsin-digestion of SILAC-labeled proteins

The main reagents and methods included: (i) Reducing agent: 100 mM DTT was added to SILAC-labeled protein sample. (ii) Uranyl acetate (UA) buffer: The UA buffer contained 8 M urea and 0.1 M Tris/HCL. The SILAC-labeled protein sample with DTT was filtered by a 10-kD ultrafiltration centrifuge tube for two times. (iii) Isolation mixture reacted: A total of 100 μL of 0.05 M iodoacetamide was added to the isolation mixture following centrifugalization (14,000 × g, 15 min). A total of 25 mM ammonium bicarbonate (NH4HCO3) was added to the mixture following centrifugalization (14,000 × g, 15 min). (iv) Trypsin buffer: 2 μg trypsin in 40 μL 100-mM NH4HCO3. (v) Tryptic peptide content: A volume of trypsin buffer (40 μL) was added to the mixture from last step and shaken by 600 rpm for 1 min. Enzymatic hydrolysis of the mixture was done for 16–18 h at 37°C. A volume (40 μL) of 25 mM NH4HCO3 was added to the mixture and that mixture was centrifuged (14,000 × g, 15 min); the filtrate was collected.

2.7 LC-MS/MS analysis

The instrument and materials are as follows: (i) MS instrument, for example, Q Exactive mass spectrometer (Thermo Fisher Scientific); (ii) Easy nLC system, for example, Proxeon Biosystems (Thermo Fisher Scientific); (iii) Thermo scientific EASY column: Acclaim PepMap, 100 μm × 2 cm, nanoViper, 5 μm-C18; (iv) analytical column: Thermo scientific EASY column (75 μm * 100 mm 3 μm-C18); (v) solvent A: 0.1% formic acid in H2O; and (vi) solvent B: 0.1% formic acid, 84% acetonitrile in H2O.

2.8 Search protein database with MaxQuant software

The main parameters are following [20]: Main search ppm: 6; missed cleavage: 2; MS/MS tolerance ppm: 20; de-isotopic: TRUE; enzyme: trypsin/P; database: uniprot_Homo_sapiens_169753_20190313; fixed modification: carbamidomethyl (C); lables: Lys(8), Arg(10); variable modification: oxidation (M), acetyl (protein N-term) decoy database; pattern: reverse; peptide FDR: 0.01; and protein FDR: 0.01.

2.9 Bioinformatics analysis

Several bioinformatics analyses were used, which are as follows: (i) The enrichment of Kyoto Encyclopedia of Genes and Genomes (KEGG) pathway was performed with R package clusterProfiler-KEGG (https://bioconductor.org/packages/release/bioc/html/clusterProfiler.html). (ii) The enrichment of biological processes

(BPs) was analyzed with Cytoscape ClueGO. (iii) The level of statistical significance was set as $p < 0.05$ and adjusted p value < 0.05. For KEGG and BPs enrichment analyses, a Benjamini-Hochberg multiple text was used to adjust p value.

3. RESULTS AND DISCUSSION

3.1 Proteomic profile based on SILAC in ivermectin-treated TOV-21G cells

The flow chart of SILAC quantitative proteomics was shown to summarize the overall analysis process for the identification of ivermectin-related proteins (Figure 1).

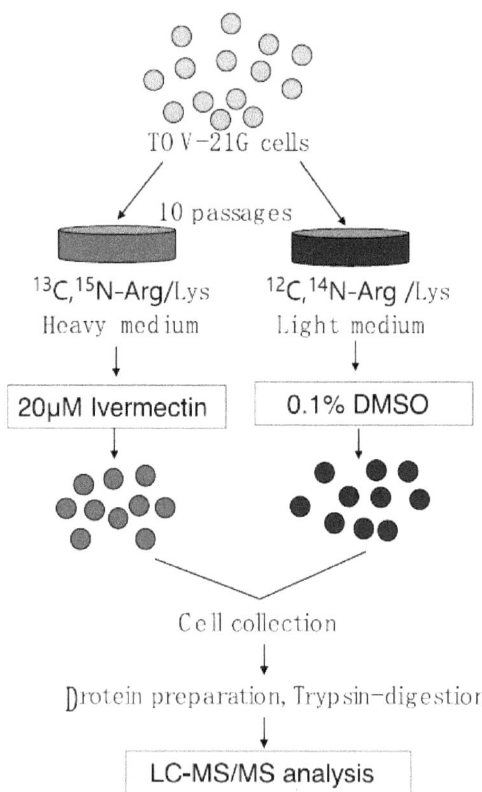

Figure 1. The flow chart of SILAC quantitative proteomics analysis of ovarian cancer cells treated with and without ivermectin.

In total, 4447 proteins were identified with SILAC quantitative proteomics in human ovarian cancer cells treated with ivermectin. The ratio of "heavy"/"light" labeling samples was obtained, including 97.91% proteins with ratio < 1, and 2.09% proteins with ratio > 1. The MS/MS spectra of tryptic peptides EYQDLLNVK (Figure 2A) and VVQGSLDSLPQAVR (Figure 2B) are taken as examples. For peptide EYQDLLNVK (gene name = NEFM), the excellent b-ion and y-ion series were obtained with high signal-to-noise (S/N) (Figure 2A). For peptide VVQGSLDSLPQAVR (gene name = PKC1), the excellent b-ion and y-ion series were also obtained with high signal-to-noise (S/N) (Figure 2B).

The fold-changes of some identified proteins were very striking; for example, those upregulated proteins (ratio > 2), including histone H2A, progranulin, cathepsin Z, SWI/SNF-related matrix-associated actin-dependent regulator of chromatin subfamily A-like protein 1, beta-mannosidase, GRAM domain-containing 1C, BMP-2-inducible protein kinase, ribosomal protein L3, ubiquitin-conjugating enzyme E2, PIK3R1, LIM domain-containing protein 1, retinal guanylyl cyclase 1, telomerase-binding protein EST1A, and COX7A2L protein. Some of them have been reported in ovarian cancers. For example, recent studies demonstrated that PI3K/AKT/mTOR and ERK1/2 signaling pathways were involved in this chemoresistance. Progranulin was upregulated in epithelial ovarian cancer cell lines and associated with cisplatin resistance through regulating AKT/mTOR and ERK1/2 signaling pathways [21]. Progranulin is also involved in the process of cartilage development, progression, wound healing, and inflammation in ovarian cancer [22]. Additionally, one study showed that progranulin could directly activate cancer-associated fibroblasts to

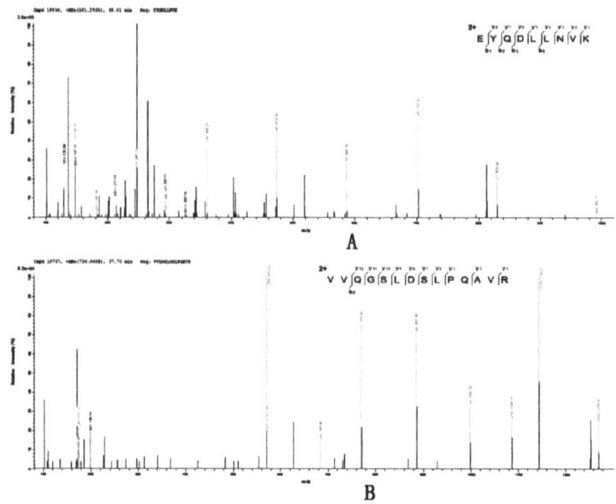

Figure 2. The MS/MS spectra of tryptic peptides with SILAC labeling. (A) MS/MS spectrum of tryptic peptide EYQDLLNVK (gene name = NEFM). (B) MS/MS spectrum of tryptic peptide VVQGSLDSLPQAVR (gene name = PKC1).

induce the epithelial-mesenchymal transition process of epithelial ovarian cancer cells [23]. Copy number loss of PIK3R1 most commonly occurs in ovarian cancer, which would activate AKT and p110-independent JAK2/STAT3 signaling and renders ovarian cancer cells vulnerable to AKT inhibitors [24]. CD97 can activate NF-κB-dependent JAK2/STAT3 pathway, consequently playing an important role in migratory, invasive capacity, and drug-resistant in ovarian cancer cells [25].

Some downregulated proteins (ratio < 0.1) were also very striking; for example, anion exchange protein, Rho guanine nucleotide exchange factor 16, dynein assembly factor 1, glucoside xylosyltransferase 1, glutamine synthetase, insulin-like growth factor-binding protein 2, microtubule-associated protein RP/EB family member 3, myelin proteolipid protein, neurochondrin, neurofilament medium polypeptide, phosphoenolpyruvate carboxykinase, ANKUB1, and TASOR 2. Some of them play an important role in the pathogenesis of ovarian cancer. For example, the most prominent effects of insulin-like growth factor-binding protein 2 in ovarian cancer include promoting driving invasion, proliferation, and suppressing apoptosis. The area under the ROC curve of insulin-like growth factor-binding protein 2 in detecting ovarian cancer was 0.815 (95% CI: 0.721–0.910, $P < 0.001$), further studies are needed to confirm its diagnostic performance at an early stage of ovarian cancer [26].

Targeting glutamine metabolism could be a new strategy to overcome cisplatin resistance in different types of cancer. Glutamine synthetase has a role in altering glutamine metabolism to cause cisplatin resistance in A2780 ovarian cancer cells [27]. Anion exchanger 2 is a chloride/bicarbonate transporter that operates independently of sodium, playing a role in controlling membrane potential and intracellular pH. Anion exchanger 2 exhibited elevated expression in ovarian cancer tissues when compared to nearby non-tumor lesions using quantitative proteomics analysis. [28]. Proteins found in ovarian cancer cells treated with and without ivermectin using SILAC quantitative proteomics revealed dependable biomarkers and therapeutic targets for ivermectin's anticancer effects [8].

3.2 Ivermectin-mediated molecular pathway in human ovarian cancer cells

In total, 89 statistically significant molecular pathways were enriched based on those 4447 ivermectin-related proteins with KEGG pathway analysis (Table 1).

Table 1. Statistically significant pathways identified with ivermectin-related proteins with KEGG pathway enrichment analysis.

Pathway ID	Pathway name	p value
hsa00010	Glycolysis/gluconeogenesis	2.03E−03
hsa00020	Citrate cycle (TCA cycle)	7.38E−11
hsa00030	Pentose phosphate pathway	3.51E−04
hsa00052	Galactose metabolism	6.04E−03
hsa00062	Fatty acid elongation	1.09E−02
hsa00071	Fatty acid degradation	1.39E−05

Pathway ID	Pathway name	p value
hsa00190	Oxidative phosphorylation	3.47E-12
hsa00230	Purine metabolism	1.08E-02
hsa00240	Pyrimidine metabolism	2.55E-03
hsa00270	Cysteine and methionine metabolism	8.93E-07
hsa00280	Valine, leucine and isoleucine degradation	2.06E-08
hsa00480	Glutathione metabolism	1.05E-03
hsa00510	N-Glycan biosynthesis	4.31E-03
hsa00513	Various types of N-glycan biosynthesis	5.15E-03
hsa00520	Amino sugar and nucleotide sugar metabolism	1.04E-07
hsa00620	Pyruvate metabolism	1.11E-07
hsa00630	Glyoxylate and dicarboxylate metabolism	4.09E-03
hsa00640	Propanoate metabolism	1.16E-05
hsa00920	Sulfur metabolism	4.45E-03
hsa01040	Biosynthesis of unsaturated fatty acids	1.09E-02
hsa01200	Carbon metabolism	3.96E-16
hsa01212	Fatty acid metabolism	1.74E-08
hsa01230	Biosynthesis of amino acids	3.09E-08
hsa03008	Ribosome biogenesis in eukaryotes	3.34E-04
hsa03010	Ribosome	7.07E-22
hsa03013	RNA transport	8.04E-20
hsa03015	mRNA surveillance pathway	4.28E-08
hsa03018	RNA degradation	2.24E-07
hsa03030	DNA replication	8.18E-09
hsa03040	Spliceosome	5.79E-24
hsa03050	Proteasome	1.50E-14
hsa03410	Base excision repair	1.22E-02
hsa03420	Nucleotide excision repair	4.72E-06
hsa03430	Mismatch repair	4.33E-04
hsa04012	ErbB signaling pathway	6.27E-03
hsa04066	HIF-1 signaling pathway	1.31E-05
hsa04071	Sphingolipid signaling pathway	9.01E-03

Pathway ID	Pathway name	p value
hsa04110	Cell cycle	4.03E-03
hsa04120	Ubiquitin mediated proteolysis	1.82E-06
hsa04130	SNARE interactions in vesicular transport	1.18E-04
hsa04137	Mitophagy—animal	9.07E-03
hsa04141	Protein processing in the endoplasmic reticulum	1.07E-13
hsa04142	Lysosome	3.81E-06
hsa04144	Endocytosis	9.24E-15
hsa04145	Phagosome	8.66E-07
hsa04152	AMPK signaling pathway	3.49E-03
hsa04210	Apoptosis	1.07E-03
hsa04213	Longevity regulating pathway—multiple species	4.46E-03
hsa04216	Ferroptosis	4.75E-04
hsa04510	Focal adhesion	2.57E-05
hsa04520	Adherens junction	2.04E-05
hsa04530	Tight junction	6.21E-05
hsa04540	Gap junction	3.14E-03
hsa04611	Platelet activation	4.03E-03
hsa04666	Fc gamma R-mediated phagocytosis	2.41E-03
hsa04714	Thermogenesis	1.08E-09
hsa04720	Long-term potentiation	7.22E-03
hsa04721	Synaptic vesicle cycle	2.29E-04
hsa04810	Regulation of actin cytoskeleton	4.94E-06
hsa04910	Insulin signaling pathway	2.90E-04
hsa04919	Thyroid hormone signaling pathway	5.20E-03
hsa04922	Glucagon signaling pathway	3.38E-04
hsa04931	Insulin resistance	1.26E-02
hsa04932	Nonalcoholic fatty liver disease (NAFLD)	1.44E-07
hsa04961	Endocrine and other factor-regulated calcium reabsorption	1.83E-03
hsa04962	Vasopressin-regulated water reabsorption	9.81E-03
hsa05010	Alzheimer disease	3.09E-07
hsa05012	Parkinson disease	5.32E-23

Pathway ID	Pathway name	p value
hsa05014	Amyotrophic lateral sclerosis (ALS)	2.55E-03
hsa05016	Huntington disease	7.93E-13
hsa05100	Bacterial invasion of epithelial cells	5.95E-10
hsa05110	*Vibrio cholerae* infection	6.22E-06
hsa05120	Epithelial cell signaling in *Helicobacter pylori* infection	1.07E-04
hsa05130	Pathogenic *Escherichia coli* infection	6.13E-09
hsa05131	Shigellosis	2.10E-07
hsa05132	Salmonella infection	7.10E-14
hsa05134	Legionellosis	1.05E-03
hsa05135	Yersinia infection	4.64E-06
hsa05163	Human cytomegalovirus infection	5.57E-05
hsa05165	Human papillomavirus infection	4.68E-03
hsa05169	Epstein-Barr virus infection	2.03E-05
hsa05170	Human immunodeficiency virus 1 infection	4.59E-08
hsa05203	Viral carcinogenesis	6.98E-05
hsa05205	Proteoglycans in cancer	2.44E-05
hsa05211	Renal cell carcinoma	2.65E-03
hsa05212	Pancreatic cancer	6.79E-03
hsa05220	Chronic myeloid leukemia	1.31E-02
hsa05230	Central carbon metabolism in cancer	1.18E-03
hsa03060	Protein export	8.73E-05

Ivermectin was found to be associated with various cancer-related molecular pathways including mismatch repair, ErbB signaling, HIF-1 signaling, cell-cycle regulation, ubiquitin-mediated proteolysis, AMPK signaling, apoptosis, ferroptosis, proteoglycans, and central carbon metabolism in cancer. Ivermectin is implicated in various cancer pathogenesis processes, including energy metabolism, immunology, stromal elements, RNA regulation, hormone signaling, and chemical manufacturing pathways. Various pathways were enriched with distinct proteins, while some pathways have proteins in common. The research indicated that ivermectin has a multifaceted impact on different signaling pathways. The results were consistent with many studies previously. Ivermectin induced PAK1-mediated cytostatic autophagy both in vitro and in vivo, which might be one of the PAK1 inhibitors and inhibits the growth of ovarian cancer, glioblastoma, breast cancer, and NF2 tumors [29]. Another investigation found that ivermectin induced apoptosis in HeLa cells by upregulating Bax and p53 expressions, enhancing cytochrome c release, decreasing the levels of CDK2, CDK6, cyclin E, and cyclin D1 [30]. The primary immunogenic features—immunogenic cell death (ICD) included the release of high-mobility-group protein B1, secreted ATP, and surface-exposed calreticulin. Recent data supported that ivermectin could kill human triple-negative breast cancer cells through mechanisms of ICD, which induced pannexin-1 channel opening and cell death [31]. Ivermectin was also reported as an RNA helicase inhibitor, which reduced precursor and mature microRNAs potentially inhibiting cell invasion and proliferation [32]. Additionally, ivermectin effectively targets angiogenesis through decreasing membrane potential, mitochondrial respiration, ATP levels, and increasing mitochondrial superoxide, and the effects proliferation, capillary network formation, and survival in human brain microvascular endothelial cell [33]. Evidence suggests that ivermectin, when combined with other medications such as daunorubicin, anti-BRAF V600 inhibitors, cytarabine, paclitaxel, and tamoxifen, shows stronger anticancer effects due to interactions between distinct signaling pathways [34]. The precise mechanisms of ivermectin are still not fully understood. The use of new medications improved individuals' health.

3.3 Ivermectin-mediated biological processes in human ovarian cancer cells

61 important biological processes were identified using GO analysis on 4447 proteins associated to ivermectin. Ivermectin was found to be associated with various cancer-related biological processes, including modulation of the canonical Wnt signaling pathway, involvement in apoptotic processes through cysteine-

type endopeptidase activity, activation of innate immune response signaling, protein targeting to membranes, modulation of T-cell receptor signaling, regulation of protein ubiquitination, activation of protein kinase activity, control of transcription by RNA polymerase II, and DNA-binding transcription factor activity. These results were in line with other earlier research. CTNNB1, also known as catenin beta 1 or IMPβ1, plays a role in the biological process of protein polyubiquitination by facilitating the nuclear entry of payloads.

Ivermectin can affect the thermal stability and α-helicity of IMPα and IMPβ1 by binding to the IMPα armadillo repeat domain [35]. CASP3 in the biological process of protein kinase regulator activity is a member of the cysteine-aspartic acid protease (caspase) family. SK-MEL-28 cells were treated with different concentrations of ivermectin (2.5 μM, 5 μM, and 10 μM). Ivermectin enhanced the apoptosis effect by the upregulation of caspase-3 activity [36]. Also, PAK1 in the biological process of protein kinase regulator activity binds to and inhibits the activity of cyclin-cyclin-dependent kinase 2 or -cyclin-dependent kinase 4 complexes, and thus functions as a regulator of cell-cycle progression at G1. Ivermectin inhibited cancer stem cells formation by regulating the binding of PAK1/Stat3 complex and the IL-6 promoter [37]. YAP1 in the biological process of positive regulation of canonical Wnt signaling pathway was involved in the development, growth, repair, and homeostasis of multiple cancers.

Ivermectin suppressed YAP1 nuclear expression and accumulation in gastric cancer cells. Additionally, in xenografts of gastric cancer cells, ivermectin inhibited tumor growth by controlling YAP1 nuclear expression [38]. The proteins discovered in ovarian cancer cells treated with and without ivermectin using the SILAC approach are crucial in several cellular signaling cascades and possess diverse biological functions. The findings offer fundamental data for future research on ivermectin in ovarian cancer.

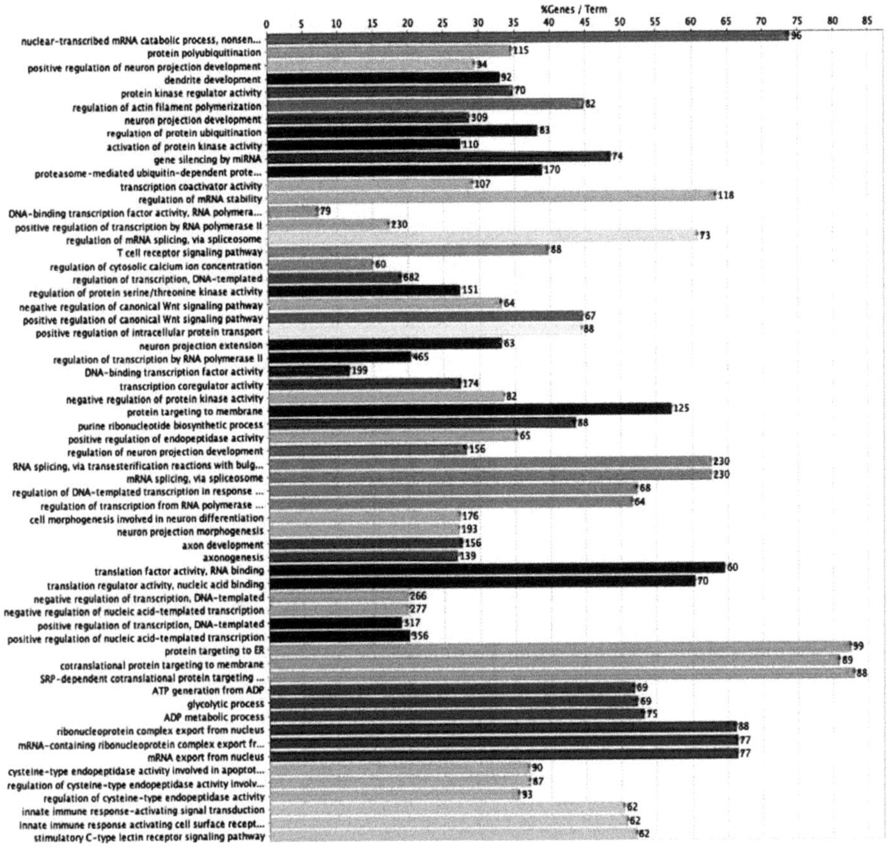

Figure 3. Statistically significant biological processes (BPs) identified with ivermectin-related proteins with GO enrichment analysis.

4. CONCISE SUMMARY

SILAC is a reliable quantitative proteomics technique used to detect variations in protein expression or modification in cultivated cells under diverse circumstances. Ovarian cancer cells TOV-21 were cultured in two different conditions using a "heavy" labeling medium containing 50 mg L-lysine-2HCl [13C6, 15N2] and 50 mg L-arginine-HCl [13C6, 15N4] in 500 mL RPMI 1640 medium, and a "light" labeling medium containing 50 mg L-lysine-2HCl [12C6, 14N2] and 50 mg L-arginine-HCl [12C6, 14N4] in 500 mL RPMI 1640 medium for 10 passages each. TOV-21G cells were labeled with SILAC "heavy" or "light" and then treated with or without 20 μM ivermectin for 24 hours. The proteins labeled with heavy and light stable isotopes were combined in identical proportions (1:1), digested using trypsin, and examined using LC-MS/MS. 4447 proteins were detected in TOV-21G cells treated with ivermectin compared to controls. These proteins showed substantial enrichment in 89 molecular pathways and 62 biological activities. The results provide crucial information for investigating the molecular pathway network alterations induced by ivermectin and identifying potential biomarkers and therapeutic targets for using ivermectin in treating ovarian cancer.

5. AT A GLANCE

SILAC involves replacing original amino acids in cell culture with isotopic essential amino acids for 8-10 generations, followed by mass spectrometry to identify and quantify proteins based on isotopic abundance differences. SILAC is utilized for profiling proteome alterations, studying protein degradation rates, examining protein-protein interactions, and monitoring dynamic variations with precise quantification and strong reproducibility. This study utilized SILAC labeling with "light" (L-Lysine-2HCl [12C6, 14N2], L-Arginine-HCl [12C6, 14N4]) or "heavy" (L-Lysine-2HCl [13C6, 15N2], L-Arginine-HCl [13C6, 15N4]) in RPMI 1640 medium to culture human ovarian cancer TOV-21G cells for 10 passages. The cells were then treated with 0.1% dimethylsulfoxide for 24 hours and 20 μM ivermectin for another 24 hours. The proteins labeled with light and heavy isotopes were combined in equal proportions (1:1) for digestion with trypsin. The tryptic peptide mixture underwent liquid chromatography fractionation and was evaluated using tandem mass spectrometry. A total of 4,447 proteins were detected in ivermectin-treated TOV-21G cells compared to the control group. The proteins were shown to be abundant in 89 major signaling pathways and 62 significant biological activities. The results showed that SILAC quantitative proteomics is a reliable method for studying how ivermectin affects the proteome of cancer cells. When combined with molecular pathway networks and biological processes enrichments, this approach offers a more comprehensive understanding of how ivermectin inhibits TOV-21G cells at a molecular level.

ACRONYMS AND ABBREVIATIONS

BPs	biological processes
CTNNB1	catenin beta 1
ICD	immunogenic cell death
DTT	dithiothreitol
EGFR	epidermal growth factor receptor
IC50	lethal concentration 50
KEGG	Kyoto Encyclopedia of Genes and Genomes
LC	liquid chromatography
MS	mass spectrometry
MS/MS	tandem mass spectrometry
OD	absorbance values
PBS	phosphate buffer solution
SILAC	stable isotope labeling with amino acids in cell culture

REFERENCES

1. Chen X, Wei S, Ji Y, Guo X, Yang F. Quantitative proteomics using SILAC: Principles, applications, and developments. Proteomics. 2015;15(18):3175-3192. DOI: 10.1002/pmic.201500108

2. Deng J, Erdjument-Bromage H, Neubert TA. Quantitative comparison of proteomes using SILAC. Current Protocols in Protein Science. 2019;95(1):e74. DOI: 10.1002/cpps.74

3. Hoedt E, Zhang G, Neubert TA. Stable isotope labeling by amino acids in cell culture (SILAC) for quantitative proteomics. Advances in Experimental Medicine and Biology. 2019;1140:531-539. DOI: 10.1007/978-3-030-15950-4_31

4. Ong SE. The expanding field of SILAC. Analytical and Bioanalytical Chemistry. 2012;404(4):967-976. DOI: 10.1007/s00216-012-5998-3

5. Stepath M, Zülch B, Maghnouj A, et al. Systematic comparison of label-free, SILAC, and TMT techniques to study early adaption toward inhibition of EGFR signaling in the colorectal cancer cell line DiFi. Journal of Proteome Research. 2020;19(2):926-937. DOI: 10.1021/acs.jproteome.9b00701

6. Kaeser-Pebernard S, Diedrich B, Dengjel J. Identification and regulation of multimeric protein complexes in autophagy via SILAC-based mass spectrometry approaches. Methods in Molecular Biology. 2019;1880:341-357. DOI: 10.1007/978-1-4939-8873-0_23

7. Tackett AJ, DeGrasse JA, Sekedat MD, et al. I-DIRT, a general method for distinguishing between specific and nonspecific protein interactions. Journal of Proteome Research. 2005;4(5):1752-1756. DOI: 10.1021/pr050225e

8. Li N, Li J, Desiderio DM, Zhan X. SILAC quantitative proteomics analysis of ivermectin-related proteomic profiling and molecular network alterations in human ovarian cancer cells. Journal of Mass Spectrometry. 2021;56(1):e4659. DOI: 10.1002/jms.4659

9. Snider J, Wang D, Bogenhagen DF, Haley JD. Pulse SILAC approaches to the measurement of cellular dynamics. Advances in Experimental Medicine and Biology. 2019;1140:575-583. DOI: 10.1007/978-3-030-15950-4_34

10. Savitski MM, Zinn N, Faelth-Savitski M, et al. Multiplexed proteome dynamics profiling reveals mechanisms controlling protein homeostasis. Cell. 2018;173(1):260-274.e25. DOI: 10.1016/j.cell.2018.02.030

11. Crump A. Ivermectin: Enigmatic multifaceted 'wonder' drug continues to surprise and exceed expectations. Journal of Antibiotics (Tokyo). 2017;70(5):495-505. DOI: 10.1038/ja.2017.11

12. Develoux M. Ivermectin. Annales de Dermatologie et de Vénéréologie. 2004;131(6-7 Pt 1):561-570. DOI: 10.1016/s0151-9638(04)93668-x

13. Levy M, Martin L, Bursztejn AC, et al. Ivermectin safety in infants and children under 15 kg treated for scabies: A multicentric observational study. The British Journal of Dermatology. 2020;182(4):1003-1006. DOI: 10.1111/bjd.18369

14. Nicolas P, Maia MF, Bassat Q, et al. Safety of oral ivermectin during pregnancy: A systematic review and meta-analysis. The Lancet Global Health. 2020;8(1):e92-e100. DOI: 10.1016/s2214-109x(19)30453-x

15. Guzzo CA, Furtek CI, Porras AG, et al. Safety, tolerability, and pharmacokinetics of escalating high doses of ivermectin in healthy adult subjects. Journal of Clinical Pharmacology. 2002;42(10):1122-1133. DOI: 10.1177/009127002401382731

16. Dou Q, Chen HN, Wang K, et al. Ivermectin Induces Cytostatic Autophagy by Blocking the PAK1/Akt Axis in Breast Cancer. Cancer Research. 2016;76(15):4457-4469. DOI: 10.1158/0008-5472.Can-15-2887

17. Sharmeen S, Skrtic M, Sukhai MA, et al. The antiparasitic agent ivermectin induces chloride-dependent membrane hyperpolarization and cell death in leukemia cells. Blood. 2010;116(18):3593-3603. DOI: 10.1182/blood-2010-01-262675

18. Li N, Zhan X. Anti-parasite drug ivermectin can suppress ovarian cancer by regulating lncRNA-EIF4A3-mRNA axes. The EPMA Journal. 2020;11(2):289-309. DOI: 10.1007/s13167-020-00209-y

19. Li N, Li H, Wang Y, Cao L, Zhan X. Quantitative proteomics revealed energy metabolism pathway alterations in human epithelial ovarian carcinoma and their regulation by the antiparasite drug ivermectin: Data interpretation in the context of 3P medicine. The EPMA Journal. 2020;11(4):661-694. DOI: 10.1007/s13167-020-00224-z

20. Trautwein-Schult A, Bartel J, Maaß S, Becher D. Metabolic labeling of clostridioides difficile proteins. Methods in

Molecular Biology. 2021;2228:271-282. DOI: 10.1007/978-1-0716-1024-4_19

21. Perez-Juarez CE, Arechavaleta-Velasco F, Zeferino-Toquero M, et al. Inhibition of PI3K/AKT/mTOR and MAPK signaling pathways decreases progranulin expression in ovarian clear cell carcinoma (OCCC) cell line: A potential biomarker for therapy response to signaling pathway inhibitors. Medical Oncology. 2019;37(1):4. DOI: 10.1007/s12032-019-1326-5

22. Yang D, Wang LL, Dong TT, et al. Progranulin promotes colorectal cancer proliferation and angiogenesis through TNFR2/Akt and ERK signaling pathways. American Journal of Cancer Research. 2015;5(10):3085-3097

23. Dong T, Yang D, Li R, et al. PGRN promotes migration and invasion of epithelial ovarian cancer cells through an epithelial mesenchymal transition program and the activation of cancer associated fibroblasts. Experimental and Molecular Pathology. 2016;100(1):17-25. DOI: 10.1016/j.yexmp.2015.11.021

24. Li X, Mak VCY, Zhou Y, et al. Deregulated Gab2 phosphorylation mediates aberrant AKT and STAT3 signaling upon PIK3R1 loss in ovarian cancer. Nature Communications. 2019;10(1):716. DOI: 10.1038/s41467-019-08574-7

25. Park GB, Kim D. MicroRNA-503-5p inhibits the CD97-mediated JAK2/STAT3 pathway in metastatic or paclitaxel-resistant ovarian cancer cells. Neoplasia. 2019;21(2):206-215. DOI: 10.1016/j.neo.2018.12.005

26. Prayudi PKA, Budiana ING, Mahayasa PD, et al. Diagnostic accuracy of serum insulin-like growth factor-binding protein 2 for ovarian cancer. International Journal of Gynecological Cancer. 2020;30(11):1762-1767. DOI: 10.1136/ijgc-2020-001479

27. Guo J, Satoh K, Tabata S, et al. Reprogramming of glutamine metabolism via glutamine synthetase silencing induces cisplatin resistance in A2780 ovarian cancer cells. BMC Cancer. 2021;21(1):174. DOI: 10.1186/s12885-021-07879-5

28. Zhang LJ, Lu R, Song YN, et al. Knockdown of anion exchanger 2 suppressed the growth of ovarian cancer cells via mTOR/p70S6K1 signaling. Scientific Reports. 2017;7(1):6362. DOI: 10.1038/s41598-017-06472-w

29. Wang K, Gao W, Dou Q, et al. Ivermectin induces PAK1-mediated cytostatic autophagy in breast cancer. Autophagy. 2016;12(12):2498-2499. DOI: 10.1080/15548627.2016.1231494

30. Song D, Liang H, Qu B, et al. Ivermectin inhibits the growth of glioma cells by inducing cell cycle arrest and apoptosis in vitro and in vivo. Journal of Cellular Biochemistry. 2019;120(1):622-633. DOI: 10.1002/jcb.27420

31. Draganov D, Gopalakrishna-Pillai S, Chen YR, et al. Modulation of P2X4/P2X7/pannexin-1 sensitivity to extracellular ATP via ivermectin induces a non-apoptotic and inflammatory form of cancer cell death. Scientific Reports. 2015;5:16222. DOI: 10.1038/srep16222

32. Yin J, Park G, Lee JE, et al. DEAD-box RNA helicase DDX23 modulates glioma malignancy via elevating miR-21 biogenesis. Brain. 2015;138(Pt 9):2553-2570. DOI: 10.1093/brain/awv167

33. Liu Y, Fang S, Sun Q, Liu B. Anthelmintic drug ivermectin inhibits angiogenesis, growth and survival of glioblastoma through inducing mitochondrial dysfunction and oxidative stress. Biochemical and Biophysical Research Communications. 2016;480(3):415-421. DOI: 10.1016/j.bbrc.2016.10.064

34. Liu J, Zhang K, Cheng L, Zhu H, Xu T. Progress in Understanding the Molecular Mechanisms Underlying the Antitumour Effects of Ivermectin. Drug Design, Development and Therapy. 2020;14:285-296. DOI: 10.2147/dddt.S237393

35. Yang SNY, Atkinson SC, Wang C, et al. The broad spectrum antiviral ivermectin targets the host nuclear transport importin α/β1 heterodimer. Antiviral Research. 2020;177:104760. DOI: 10.1016/j.antiviral.2020.104760

36. Deng F, Xu Q, Long J, Xie H. Suppressing ROS-TFE3-dependent autophagy enhances ivermectin-induced apoptosis in human melanoma cells. Journal of Cellular Biochemistry. 2018. DOI: 10.1002/jcb.27490

37. Kim JH, Choi HS, Kim SL, Lee DS. The PAK1-Stat3 signaling pathway activates IL-6 gene transcription and human breast cancer stem cell formation. Cancers (Basel). 2019;11(10):1527. DOI: 10.3390/cancers11101527

38. Nambara S, Masuda T, Nishio M, et al. Antitumor effects of the antiparasitic agent ivermectin via inhibition of Yes-associated protein 1 expression in gastric cancer. Oncotarget. 2017;8(64):107666-107677. DOI: 10.18632/oncotarget.22587

CHAPTER 6

EXPLORING THE POTENTIAL OF NANOTECHNOLOGY IN MAMMALIAN CELL CULTURE AND ITS IMPLICATIONS FOR INTELLECTUAL PROPERTY RIGHTS

1. INTRODUCTION: AN OVERVIEW

Nanomaterials are synthetic chemical compounds or materials with a particle size ranging from 1 to 100 nanometers in diameter. Presently, there is extensive research and application of nanomaterials in numerous industries for commercial purposes. Advanced nanomaterials have shown considerable promise in the fields of biotechnology and biomedicine [1]. Nanomaterials possess particular physical and chemical properties due to their small size within the nanoscale range, significant surface area, modifiable surface charge, distinctive composition, varied morphologies, and surface composition. Nanomaterials (NMs) possess distinct physicochemical characteristics in comparison to their larger counterparts, leading to improved performance, sensitivity, efficiency, and selectivity in the advancement of biomedicines. Various nanomaterials, including metal nanoparticles, liposomes, quantum dots, polymeric micelles, dendrimers, and carbon-based nanoparticles, are evaluated for their biological efficacy in nanomedicine applications. Passive targeting and active targeting are two essential ways for delivering drug-loaded nanoparticles to the specific parts of the body that are affected. Passive targeting is achieved by enhancing the permeability and retention (EPR) process. The inactive targeting strategy utilizes surface-functionalized nanoparticles with distinct biomarkers that bind to receptors that are excessively expressed in the sick tissue[3]. The importance of progress in cell culture in the medical domain has been recognized for a substantial duration. Mammalian cell culture entails the isolation of cells from a specific organ tissue and their subsequent cultivation in a suitable synthetic environment. Multiple methods of disaggregation can be employed to produce the first separation of cells from the designated organ tissues. Primary cells are often derived from living organisms, although some cells are acquired as established cell lines. MCCs are frequently employed in the biomedical industry to investigate several applications.The user's text is "[4]." Researchers highly appreciate cell culture-based investigations due to their regular and dependable outcomes, making them an essential model system in the field of cellular and molecular biology. MCC require an ideal environment for their growth, encompassing both favorable nutritional and physicochemical parameters. The importance of progress in cell culture in the medical domain has been recognized for a substantial duration. Mammalian cell culture entails the isolation of cells from a specific organ tissue and their subsequent cultivation in a suitable synthetic environment. Multiple methods of disaggregation can be employed to produce the first separation of cells from the designated organ tissues. Primary cells are often derived from a living organism, although some cells are obtained as established cell lines. MCCs are frequently employed in the biomedical industry to investigate several applications. The user's text is "[4]." Researchers highly appreciate cell culture-based investigations due to their constant and dependable outcomes, making them an essential model system in the field of cellular and molecular biology. MCC require an ideal environment for their growth, encompassing both favorable nutritional and physicochemical parameters. The user's text is "[5]." Animal, plant, and bacterial cells are grown in a carefully regulated environment using a specialized culture medium. Animal cell cultures are more complex than plant and bacterial cell cultures due to their genetic composition. Overseeing the growth of adult stem cells and cultivating pluripotent stem cells is a challenging endeavor. The progress in stem cell culture technology has significantly contributed to the successful development of 3D organoids that accurately mimic tissues. The fields of nanotechnology (NT) and related sciences have experienced significant growth in biomedical applications, including diagnostics, testing, tracking, medication delivery, nanomedicine, medical implants, and electronics. This is due to their compatibility with biological entities. Biomedicine encompasses the creation and production of nanomaterials (NMs), as well as other nanoparticles (NPs) and nanodevices [7]. Once appropriately constructed, nanoparticles (NMs) demonstrate their inherent ability to circulate through the bloodstream along several pathways determined by their characteristics, ultimately reaching all organs. Because of their inherent biocompatible interactions, the nanoparticles (NPs) possess distinct physicochemical characteristics that are linked to reduced immunogenicity and non-toxicity. Using NMs for biological applications offers several advantages. Firstly, it allows for increased drug concentration in pathological tissues and enables controlled release of the drug. Secondly, it addresses problems related to low solubility and bioavailability of the drug. Lastly, it enhances biodegradability and biocompatibility. Drugs, genes, and imaging agents can be readily loaded onto the surface of the material because of its adjustable surface functions [1, 7, 8]. Imaging agents have the potential to enable the monitoring of medications in the body, allowing for the evaluation of how effectively the pharmaceuticals are distributed throughout treatment. Recent progress has resulted in the creation of several kinds of nanoparticles, such as liposomes, polymers, metal nanoparticles, and inorganic nanoparticles. These nanoparticles are specifically engineered to selectively attack tumor cells and other diseased tissues, while minimizing harm to healthy cells and organs. This chapter will examine the application of nanotechnology and Intellectual Property Rights (IPR) in mammalian cell culture.

2. COMPATIBILITY OF NANOMATERIALS TOWARDS BIOLOGICAL INTERACTIONS

Nanomaterials (NMs) are fascinating because to their unique, customizable, and versatile physicochemical properties, straightforward manufacturing methods, compatibility with living beings, and capacity to alter their surfaces [1]. Nevertheless, the compatibility of nanoparticles with biological entities is a vital phenomenon that emphasizes the necessity for in-depth research [9]. The initiation of most bio-applications, such as drug administration, bioimaging, and treatment, involves the binding of nanoparticles to the specific cells being targeted. The biocompatibility of nanoparticles is determined by various physical and chemical characteristics, such as diameter, shape, composition, concentration, functionalized moieties, and surface potential (Figure 1) [10]. Quantum dots have emerged as a novel bio-imaging tool among several types of nanomaterials, thanks to their distinct and adjustable physicochemical characteristics. Prior studies have influenced the creation of adaptable quantum dots that exhibit strong fluorescence and remain stable in various biological conditions. In addition, researchers have created quantum dots that are surrounded by amphipathic polymers. These quantum dots have also been modified on their surface with receptor targeting ligands. These modified quantum dots have shown promise for use in bio-imaging and drug-delivery experiments conducted on animals. Remarkably, these materials were discovered to be compatible with the cells. Nevertheless, it is necessary to evaluate their comprehensive long-term genetic toxicity, as well as their compatibility with blood and organs [1, 7, 11].Polymeric nanoparticles are highly esteemed in the fields of drug and gene delivery, tissue engineering, and numerous biomedical applications due to their non-toxic characteristics and their compatibility with biological systems. These particles are colloidal in nature and are composed of polymers that can be either natural, synthetic, or semi-synthetic. In this regard, biodegradable nanoparticles produced from a highly compatible triblock copolymer are used for non-viral gene transfections [3]. Liposomes are a widely used type of nanomaterial drug delivery technology that has been extensively studied and modified due to their favorable physicochemical features that are compatible with biological systems [12]. Liposomes consist of one or many layers of lipids surrounding a central aqueous core. The nanoscale carrier system offers substantial advantages, including: biodegradability, biocompatibility, ease of synthesis, less toxicity, sustained drug release, and the ability to incorporate hydrophilic and hydrophobic drugs. Liposomal-surface modification is a crucial strategy for targeted therapy and especially for cancer treatment [13]. Seventeen liposomal formulations have received clinical approval for cancer, inflammation, infectious disorders, antibiotic medicines, and anesthetics, with several others currently undergoing different rounds of clinical testing.[12, 13].

Although there have been promising biomedical applications of nanoparticles, the comprehensive examination of their biocompatibility, including thorough studies of acute and chronic toxicity, is not well understood. In addition, the objective of the nanoparticle design is to determine the advantageous physicochemical characteristics of various materials by assessing their toxicity. Therefore, the bio-molecule that is actively engaged with biological entities should have a strong affinity for nanoparticles that come into direct touch with biological objects, rather than their temporary initial dispersal. Several types of nanoparticles, including liposomes, lipoplexes, polymeric nanoparticles, polyplexes, metal nanoparticles, metal oxides, dendrimers, and quantum dots, have been carefully designed for their use in medical applications such as diagnosis, drug and gene delivery, tissue engineering, and biosensing [8, 13, 14]. Moreover, it is crucial to thoroughly assess and examine the compatibility and potential toxicity of nanoparticles in order to attain therapeutic efficacy. This section will explore how the physical and chemical properties of engineered nanoparticles can be altered to achieve specific biological objectives without posing any potential risks.

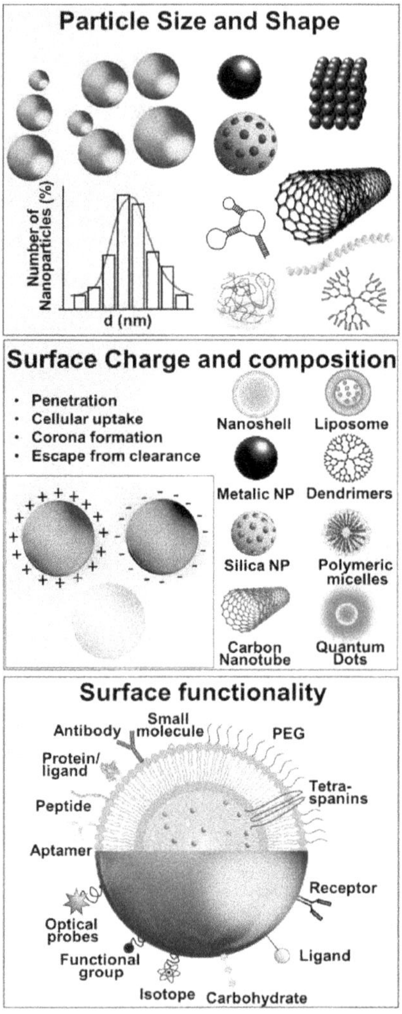

Figure 1. Precision of targeted drug delivery using nanocarriers and bio-compatibility. Nanoparticle based drug delivery platform depends on surface functionality, size and shape and surface charge and composition.

2.1 Adjustable physical and chemical properties of nanoparticles suitable for medical uses

Nanoparticles possess exceptional physical and chemical characteristics that can be controlled to maximize their advantages, such as their adjustable size, large surface area, diverse shapes, variable concentrations and compositions, and surface modifications. [15] (Figure 1). Interactions of NMs to the cell surface, their internalization, and subcellular localization, communication with the cells eventually contribute to therapeutic or adverse effects. Understanding the physicochemical attributes of NMs and their interactions with biological entities can help design superior NMs for further applications. We are jotting down the relevant physicochemical attributes of nanoparticles, which may modulate their function in therapeutic or toxicity aspects; thus, they need to be engineered wisely [16].

2.1.1 Nanomaterials size

The most important aspect of manufactured nanoparticles is their size, which influences other physicochemical properties. The smaller size of the nanoparticles allows for increased cellular localization, enhancing their interaction with cellular tissues, particularly diseased tissue, to achieve a specific biological effect for therapeutic purposes. Bio-distribution studies were conducted to analyze the impact of size variation on drug-conjugated silica nanoparticles of three different sizes (20 nm, 50 nm, and 200 nm). Nanoparticles with a diameter of 50 nm showed the greatest tumor targeting, improved retention in cancer tissue, and delayed elimination. [16]. Moreover, nano-sized particles preside over their pharmacokinetics, are predictable to traverse biological barriers, which is not possible for bulk particles. Besides, ~50 nm diameter particles showed higher efficacy because of active engagement to the biological tissues, modulating pathways, and cellular activities [17].

2.1.2 Nanomaterials surface charge

The surface charge is a unique character of NMs to manage its therapeutic and toxicological effects and plays a significant role in electrostatic interactions of NMs and living entities (Figure 1) [10]. Besides, the cellular localization pathways and tissue interactions are regulated by the surface charge of the nanoparticles, thus playing a significant role in the compatibility and cellular toxicity. Several reports suggest that nanoparticles with a positive charge highly interact with the negatively charged cell membranes and provoke genotoxicity [18]. Positively charged cationic liposomal drug and gene delivery systems have been extensively studied for the last decade. It was recently shown that cationic lipoplexes are not showing any genotoxicological aberrations in the Swiss albino mice. Typically, cell membranes are anionic in nature; thus, negatively charged NMs have very slow cellular internalization compared to neutral and positive nanoparticles [14]. Surface potentials of metal particles in controlling various tumorous and non-tumorous tissue types have been determined. Multiple studies have indicated the significance of surface potential in various nanoparticles and their interactions with biological entities. Surface charge plays a crucial role in modifying the biological functions of nanoparticles, enabling the design and engineering of nanoparticles for specific cellular targets in different diseases with reduced toxicity [16, 18].

2.1.3 Surface functionalization

Nanoparticles are crucial for enhancing the intracellular transport of therapeutic drugs and improving their retention in diseased cells over healthy ones. [1]. Surface functionalization with suitable receptor-targeted ligands using different methods results in the formation of targeted nanoparticles with improved therapeutic response and minimized off-target side effects by prolonging their circulation time in blood, increasing target specificity, cellular uptake, and drug accumulation in the tumors (escaping lysosomal degradation and enhancing stimuli-responsive drug release)

(Figure 1) [17, 19]. Depending on their application, nanoparticles are functionalized with different targeting ligands either by directly conjugating ligands to PEGylated nanoparticles through post-insertion technique or by covalent grafting on the surface of the nanoparticles. In this context, surface functionalization of nanoparticles with antibodies, peptides, folic acid, aptamers has been extensively studied. This prompts scientists to design and engineer nanoparticles for selective targeting and high retention in the tumor tissue rendering minimal toxicity to the vital organs [19, 20].

3. MECHANISM OF TARGETED DRUG DELIVERY USING NANO-CARRIER

Nanoparticles are crucial for enhancing the intracellular transport of therapeutic drugs and improving their retention in various diseased tissues compared to alternative treatments. [21]. Like normal tissues, tumors need nourishments by means of food and oxygen and a capacity to remove metabolic excretes and carbon dioxide. Diverse patterns of tumor-associated neovascularization, obtained by angiogenesis, cope with these demands. Primary conservative treatment modalities involved in cancer treatment are surgery, radiotherapy, and chemotherapy, while additional therapies such as immune therapy, targeted therapy, and hormone therapy are chosen depending on the type of tumor [22]. Nanoparticles are crucial for enhancing the intracellular transport of therapeutic drugs and improving their retention in various diseased tissues compared to alternative treatments.

3.1 Nanoparticle drug delivery by passive targeting

Passive targeted drug delivery mainly depends on the physicochemical attributes of the NMs, such as shape, diameter, surface potentials, and pathophysiological conditions of the disease microenvironment. Intravenously injected drug encapsulated NMs tend to disperse throughout the body evenly [23]. However, unlike normal tissues, tumor cells tend to take up particles of a definite diameter to a greater extent than healthy cells due to the arrangement of capillary endothelial cells, accumulating extravasated molecules in the interstitial spaces poor lymphatic drainage increases the permeation and accumulation of drug-mediated NMs. This type of NMs accumulation in the tumor region is known as the EPR effect [1, 2]. The EPR effect is influenced by physicochemical attributes of NM including particle diameter, shape, and surface potentials greatly influence the circulation time, penetration speed, tumor localization, and intracellular internalization.

Particle diameter plays a critical role in achieving effective drug delivery as it enhances permeation and circulation time and reduces renal clearance. For example, phagocyte cells facilitate larger particle uptake, while non-phagocytic cells favor the uptake of smaller particles. PEGylated NPs reduced plasma protein adsorption on their surface and reduced hepatic filtration when their size is smaller than 100 nm [24]. Particle diameter with 20–200 nm effectively enhances the permeation in both hyper-permeable and poorly permeable tumors, and particles with less than 6 nm avoid renal clearance. The NPs surface potentials could play a vital role in circulation and cellular localization [24, 25]. NPs with positive surface potentials such as cationic liposomes induce non-specific interactions with blood components and aggregation of liposomes results in a reduction of EPR effect and increased renal clearance. However, positively charged NPs are more readily taken up by cancer cells. Whereas anionic and neutral surface potential-bearing NPs circulate longer in the blood circulation [1, 2, 24].

Besides, Polyethylene glycol (PEG) polymer is used as a stabilizer (stealth liposomes) that increases the circulation time in blood up to 24–48 hours and improves in vivo stability [26]. PEG-coated liposomes induce the 'steric stabilization effect' by creating hydrophilicity on the surface of liposomes that shield surface charge and increases the repulsive forces between liposomes and blood components. Thus, it prevents aggregation of liposomes and opsonization by the reticuloendothelial system, macrophages, mononuclear phagocytic cells and prolongs their systemic circulation. On the other hand, PEG-coated liposomes induce PEG-specific IgM antibodies, enhancing hepatic uptake and rapid clearance of liposomes from systemic circulation on subsequent administration. PEG corona produces steric hindrance with tumor cells that prevent effective internalization, which could be minimized by using short PEG chains with molecular weight less than 1000 Da or by designing PEG with enzyme-cleavable bound or tumor-targeting ligands [20, 26]. To investigate the influence of shape on the cellular localization of NPs, Li et al. conducted large-scale molecular simulations to evaluate different NP geometries with identical surface area, ligand-receptor interaction strength, and PEG grafting density. They observed that spheres exhibited the fastest internalization rate, followed by cubes, while rods and disks were the slowest. Many liposomal formulations have received clinical approval, like Doxil, Abraxane, etc. However, nanoparticles grafted with PEG prolong the systemic circulation of the particles and induces the EPR effect in tumor cells, but lack of target specificity often results in reduced therapeutic efficacy [27]. Because of that, more than 95% of passively targeted formulations fail to go bench to bedside.

3.2 Nanoparticle-based drug delivery by active targeting

An ideal nanoparticle delivery system should be proficient at reaching, recognizing, and delivering its payload to determined morbid tissues and avoid drug-induced toxicity to healthy tissues [7]. Therefore, functionalizing specific targeting moieties on the surface of nanoparticles is the most usual plan. Nanoparticles are functionalized on their outer surface by targeting moieties such as small molecule ligands, monoclonal antibodies, aptamers, cell-penetrating peptides, and proteins that are internalized into morbid cells by interacting with cell surface receptors like folate receptors, transferrin receptors, tyrosine kinases like EGFR, and so on [28] (Figure 1). Cell surface receptors that are significantly overexpressed in diseased cells, compared to normal healthy cells, provide a potential target for the design and development of actively targeted drug delivery and help to reduce off-target effects [7, 17]. These ligand moieties can interact with target-specific diseased cells and protect nanoparticles from enzymatic demolition.

Targeted drug delivery significantly minimizes the toxicity and induces patient compliance with less frequent dosing. Active targeting depends on ligands bound to the NP surface to improve their uptake selectivity and protect NPS from enzymatic destruction. The main principle of active targeting involves functionalizing an NP with a ligand that binds to a molecule overexpressed on cells. Ligands with a high binding affinity to a specific cell type exhibit higher delivery efficiency. One important thing to consider is that healthy cells still express the same molecule, and as healthy cells greatly outnumber, the chances of NPs missing their target will also increase. An intelligent selection and functionalization with multiple ligands can effectively mitigate the problem. Apart from this, active targeting mainly determined the kind of nanoparticle carrier, ligand targeting specific receptors, functional agents used for linking a ligand to the nanoparticles, hydrophilic polymers, and encapsulated active ingredients [28, 29].

Targeting tumor cell surface receptors is a common approach in active targeting. Nanoparticles were linked with targeted ligands for targeting specific cell receptors and thus upregulated the intracellular localization and therapeutic efficiency. Liposomes are conjugated with antibodies, a Y-shaped glycoprotein, or its fragments often termed as immunoliposomes, increasing the specificity of liposomes by targeting antigen-presenting cancer cells, which undergo endocytosis and destroy cancer cells followed by immune system clearance [28]. Folate receptors are membrane proteins overexpressed by various tumor cells. Folic acid is a ligand for targeting folate receptors, which pose high affinity, stability, and conjugation capacity [30]. It is conjugated with nanoparticles and a PEG spacer that inhibits steric hindrance between the cells and liposomes, which helps to increase cellular uptake and drug delivery of folate-targeted anticancer drugs. Targeting folate receptors with folic acid ligands helps deliver therapeutic and imaging agents effectively to the requisite site. Endothelial growth factor receptors (EGFR) overexpressed in solid tumors like non-small cell lung cancer, colorectal, squamous cell carcinoma of the ovary, kidney, head, neck, pancreas, prostate, and breast cancers can help in designing EGFR targeted drug delivery system. Antibody fragments used for targeting EGFR are functionalized on nanoparticle surfaces in order to acquire high targeting specificity [31]. Fibroblast growth factor receptors are overexpressed in cancers like lung, prostate, bladder, etc. Several groups have reported remarkable interaction of FGFs conjugated liposome with FGFR and discussed in detail [32, 33]. Overexpression of CD44 is observed in cancers like leukemia, ovarian, colon, gastric, pancreatic, and epithelial cancers. Hyaluronic acid acts as a ligand for CD44 and is used to deliver gemcitabine and DOX encapsulated within the liposomes [34].

Targeting the tumor microenvironment is another approach in active targeting, and one aspect is targeting the tumor vasculature instead of the tumor. This approach helps in the targeted destruction of neo-angiogenic blood vessels essential for tumor growth and metastasis [29, 35]. Vascular endothelial growth factor receptors (VEGFR) play a significant role in tumor angiogenesis and vascular permeability and regulate other aspects of tumorigenesis. Bevacizumab, a monoclonal antibody approved by USFDA, is used as an anti-human VEGF for targeting VEGFRs and FGFRs tyrosine receptors for active targeting [29]. Vascular cell adhesion molecules (VCAM-1) are cell adhesion molecules (CAMs) present on the endothelial cells responsible for inflammation. VCAM-1 is overexpressed in cancers like non-small cell lung cancer and tumor vasculature. Anti-VCAM and Fab-conjugated liposomes have high cellular uptake into Human Umbilical Vein and Endothelial Cells (HUVEC) compared to conventional liposomes [36].

Matrix metalloproteases (MMPs) are calcium-dependent endopeptidases involved in remodeling extracellular matrix, tumor invasiveness, and metastasis by modulating the formation of new blood vessels [37]. Conjugating MMP-2 cleavable peptides to liposomes loaded with cell-penetrating peptides increase the tumor selectivity. αβ-integrins are the heterodimeric transmembrane glycoproteins that facilitate the adhesion of endothelial cells with adjacent tissue and blood vessels. A tripeptide Arg-Gly-Asp (RGD) exhibited high specificity for αvβ3 integrin helps in developing integrin targeted liposomes, which inhibits adhesion and angiogenesis in the tumor microenvironment (TME) [38]. Active targeting modifies the inherent behaviors of a nanocarrier to guide it towards the particular diseased tissue. Passive targeted delivery relies on the spontaneous distribution of therapeutic motifs and the enhanced permeability and retention (EPR) effect. Both targeting strategies rely on blood circulation and the site of initial medication administration. Nevertheless, there are few instances of commercial progress achieved by the use of actively targeted nanoparticles [39].

4. NANOMATERIAL AND THEIR APPLICATION FROM BIOLOGICAL ANALYSIS

4.1 Nanomaterial-driven faster and more accurate cell analysis

Timely identification and diagnosis are crucial in combating various diseases. Scientists use the distinctive properties of nanomaterials to create new molecular contrast agents for in vivo imaging, sensing, monitoring therapy response, and liquid biopsy for studying illness development, advancement, and treatment effectiveness. Nanotechnology offers a wide range of precise cell analyses.Nanotechnologyenablesthecreationofuniqueformulations for analyzing individual cells and providing customized treatments. It offers novel applications for cell sensing, imaging, administration, and diagnosis. [39]. Since the importance of accurate cell analysis for nanoparticles is the latest approach, there is a big void for more discoveries and optimizations in various bio-applications.

4.2 Nanomaterial and in vivo imaging

The main lacunae in cancer treatment are a late diagnosis. The resolution of current imaging methods is low and can detect cancers at the late/ advanced stage or metastasized. A tissue biopsy can only help physicians to ascertain the tumor type and characteristics. Detection becomes even more challenging when metastatic modules and micrometastasis need to be identified. In vivo imaging enables us to non or minimally-invasively delve deep into the patient's tissue and is becoming increasingly popular for basic research and clinical applications. In vivo, molecular imaging focuses on obtaining spatiotemporal information about molecules of medical interest or biomarkers within a living body in real-time. Molecular in vivo imaging relies on contrast agents or medium that increases the contrast of physiological structure and enhances the sensitivity of detection. Different contrast agents are used for different in vivo imaging techniques including, radiocontrast, magnetic resonance imaging (MRI) contrast, ultrasound contrast, and optical contrast agents [40]. Precision diagnostics is dependent on high-resolution and high-contrast images. Nanomaterials are critical players in the generation of advanced contrast agents or media. Imageable nanoparticles can be classified based on their applications in nuclear, magnetic, optical, and acoustic imaging modalities. Moreover, NP-based contrast agents may be designed to integrate multiple detection modules and target specific cells. The advantages of nanoparticle-based contrast agents include enhanced specificity, increased photo and chemical stability, longer circulation time, engineered clearance pathways, and multimodal applications. The main in vivo imaging modalities include MRI, computed tomography (CT), positron emission tomography (PET), single-photon emission computed tomography (SPECT), ultrasonography (US), near-infrared fluorescence (NIRF), and two-photon intravital microscopy [41, 42, 43].

4.3 Nanoparticles as bio-sensors

Nanoparticles are suited for nano bio-sensing applications due to their unique features, which provide improved sensitivity. Nanoparticles have broad applications in cell and pathogen detection, pathogen separation, identification of various biological substances, and detection of molecular and cellular activities. [41, 42]. Accurate and professional separation of desired cells from the composite of various cell mixtures is essential for numerous biological applications. Nanoparticles have been investigated as a promising and very sensitive tool for the specific identification of cells. Identification and incarceration of metastatic cancer cells in the circulation can help understand and a strong analytical biomarker for various metastatic cancers, which can change the patient's prognosis. Nanoparticle-based methods are more frequently used for the identification and capture of metastatic circulating cancer cells. In this technique, magnetic nanoparticles were used to specifically track and separate the cells by using a ligand-receptor-based mechanism [42]. These techniques can also be used for the white blood cells with an anti-CD45-APC as a nanoparticle targeting ligand [44].

Additionally, various nanoparticle-based technologies have been investigated as a sensor for the identification and selection of various pathogens. The most frequently used method for finding bacteria is magnetic biosensors that involve immunological mechanisms using magnetic nanoparticles functionalized with antibodies against surface antigens. Many researchers have been utilizing small molecule tethered nanoparticles to analyze the bacteria successfully. Magnetic glyco-nanoparticles mediated particles could detect bacteria within 5 minutes, including subtraction from the sample by the bacterial interaction with carbohydrates on mammalian cell surfaces [41].

4.4 Nanoparticles as imaging agents

Nanoparticles have been studied as imaging agents because of their outstanding physical and chemical properties for several medical uses like cancer and heart problems. Fluorescent labels can be readily attached to nanoparticle surfaces using several chemical techniques to create a diverse array of imaging agents for real-time cellular imaging in laboratory settings and living organisms. [45, 46].

Due to their passive and active targeting nature, nanoparticles can easily identify their specific biomarkers and accumulate at high concentrations in the targeted tissue. The high capability for nanoparticle modification and retention properties in the specific tissue region empowers their utilization as imaging amplifiers. Quantum dots are the most promising fluorescent labels for cellular imaging among all nanoparticles due to their inherent near infra region light emitting nature, reducing autofluorescence [47].

RGD peptide conjugated self-emitting quantum dots can be used for specific integrins highly expressed in tumors. The targeted nanoparticle has been examined for complex imaging competence, like imaging various molecular targets using different spectral emissions specific nanoparticles. Recently, nanotechnology has been used for imaging metastatic tumor cells in circulation, tumor cells, and their vasculature, stem cells, and lymph nodes [48]. Che et al. designed shortwave infrared window (SWIR)-responsive QDs for bone-specific real-time in vivo and ex vivo imaging and could visualize the significant bone structures Balb/C nude and Balb/C mouse [49]. The use of specific nanoparticles can help accurately decipher and image the gram-negative and gram-positive bacteria. Due to their fluorescence characteristics and specific bacterial cell wall interactions, they can be used in a wash-free fashion in bacterial imaging, which is significant for health care, food processing, and medical hygiene.

4.5 Application of nanoparticles in theranostics

Theranostic NMs are designed by the consolidation of diagnostic and therapeutic abilities in one biodegradable nanoparticle [50]. Novel theranostic materials should have the following properties; i) highly compatible with biological entities, ii) proficiently and precisely accumulate in desired morbid tissue, iii) describe the biochemical and morphological attributes of maladies, iv) exhibit minimal toxicological effects, v) and deliver a sufficient amount of therapeutic agent. Several techniques have been used to functionalize the surface of nanoparticles for theranostics use. Surface functionalization may include imaging agents, drugs, therapeutic cargo, nucleic acid, and contrast agents by either chemical functionalization or by biofunctionalization. Chemical functionalization depends on chemical cross-linking, while biofunctionalization of nanoparticles relies on bioinspired ligands obtained from natural phytochemicals). The use of nanotechnology offers a promising alternative for the diagnosis of various cancers. Various investigations convey that nanoparticles could be engineered for advanced diagnostic agents to detect cancers [51]. Double drug encapsulated liposomes can be functionalized to enhance theranostic efficacy [51, 52]. Multifunctional Metal nanoparticles can serve as a unique platform for cancer theranostics. The range of use of metal nanoparticles includes MRI imaging, biological catalysis, magnetic hyperthermia, magnetic drug delivery, photo-responsive drug delivery, and cell separation. Metal nanoparticles, including, Polymer-NP constructs containing $Gd3+$ complexes, $Fe3 + -$ terpyridine complexes, and polymeric shell-based contrast agents, are widely studied for their theranostic use as MRI contrast. Magnetic particle imaging (MPI), a novel imaging technique, is based on the analysis of iron oxide NPs in response to a magnetic field.

Cheng et al. used GE11, a novel peptide with EGFR binding affinity and complexed with doxorubicin-loaded liposomes, and observed higher liposomal uptake and accumulation than, unconjugated liposomes using NIRF [53]. In another study Song et al. designed a multifunctional targeting liposome for targeting lung cancer. Octreotide (OCT), a synthetic 8-peptide analog of somatostatin, was used to surface coat the liposome for enhanced binding with the somatostatin receptors overexpressed in a subset of tumors. Double anti-cancer drug (Honokiol and epirubicin) co-encapsulated liposomes showed enhanced OCT- somatostatin receptor binding and in vivo response [54]. Cittadino et al. designed a theranostic long-circulating liposome with co-loaded prednisolone phosphate and an amphiphilic paramagnetic gadolinium contrast agent

[Gd-DOTAMA(C18)(2)] for MRI monitoring of melanoma. The theranostically engineered liposomes showed long-term MRI-based detection without a loss in drug action [51]. The theranostic nanoparticle could assist in the patient's pre-selection, a prediction for responding to nanomedicine therapy. Moreover, nanomedicine-treated patients could be monitored throughout treatment duration while using nanomedicine formulations [39].

5. BIOSAFETY AND BIOETHICS ISSUES IN HANDLING MAMMALIAN CELLS

Biosafety is a notion that requires protecting human health and the surroundings of pathogenic and genetically modified mammalian cells or organisms used in the research. Mammalian cell culture is identified as a shelter for infectious etiologic substances, and it should change the compliance with containment measures recommended for the etiologic agent itself. The utility of cell cultures comes under the preview of a range of regulatory provisions that consider the estimation of biological risks. Genetically modified mammalian cell cultures were used in different continents; in that case, a bio-safety assessment should be regulated. The major guidelines issued to mitigate the biological risks for the users and environment are mainly by the World Health Organization; the Centers for Disease Control and Prevention, and the Swiss Expert Committee for Biosafety. Several countries or geographical zones have different directives; for example, in Europe, genetically modified research was brought into the regulatory provision (Directive 2009/41/EC). Mammalian cell culturing activities focusing on developing pharmaceutical drugs are covered by the Regulation (EC) No 726/2004 and its amendment laying down actions for the authorization and direction of medicinal goods for human and animal use. 3D cultures, especially organoid culture systems, are regularly used for disease modeling and studying nanomaterial-based physiological effects. Human Pluripotent stem cell-derived organoids are being generated from various human cell types and need better bio-safety and bioethics assessment. It must be ascertained that rules focusing on extenuating the biological risks for laboratory researchers, public health, and the environment falls under the preview one or several regulatory provisions based on biological risk assessment. Here, we are going to address the bio-safety issues involving mammalian cell cultures.

5.1 Bio-safety assessments of mammalian cell cultures

Biosafety refers to the way of protecting scientists, the health of other humans, and the environment from the probable side effects of microorganism, pathogenic, and genetically modified organisms and cells from human and mouse backgrounds. Laboratory biosafety uses safety principles and techniques to minimize the health hazard from accidental exposure or unplanned spillage while using infectious agents, toxins and other biological hazards in the laboratory setting. The bio-safety assessments applied to mammalian cells depend on a systematic assessment of the intrinsic attributes of the mammalian cultures like genetically modified cells and contaminated or intentionally infected with pathogens. Figure 2 shows a summary of the biosafety assessment and management process that is followed while handling cell culture-based experiments. This also considers an exposure analysis, which means that type of exploitation carried out with the cultures should be considered. The risk analysis of cell cultures that carry the pathogens follows the same methods for analyzing pathogens themselves. Primarily, the inclusive depiction of major pathogens is measured by the subsequent guidelines (i) pathogenicity and the infectious dose (ii) mode of transmission, (iii) host range, (iv) the epidemiology, potential reservoir and vectors, and the ability to zoonosis (v) the stability and the resilience of the pathogens in the surroundings.

Moreover, information related to the physicochemical properties of the pathogenic organism is considered, such as (i) susceptibility to disinfectants, (ii) physical inactivation, and (iii) drug susceptibility (e.g., sensitivity and known resistance to antibiotics or antiviral compounds). Lastly, aspects related to the disease caused by the pathogen are also to be taken into consideration. This includes (i) the availability of effective prophylaxis, (ii) the availability of efficient therapy, and (iii) any reported case of laboratory-acquired infections (LAIs). Even though underemphasized, several LAIs of mammalian cell cultures (or having virus suspension) has appeared. Among all, the exposure to vaccinia viruses amplified in mammalian cell cultures causes infections to laboratory researchers. Guidelines have been developed recently to work cautiously with vaccinia viruses and take a count of LAIs relating to this virus [55].

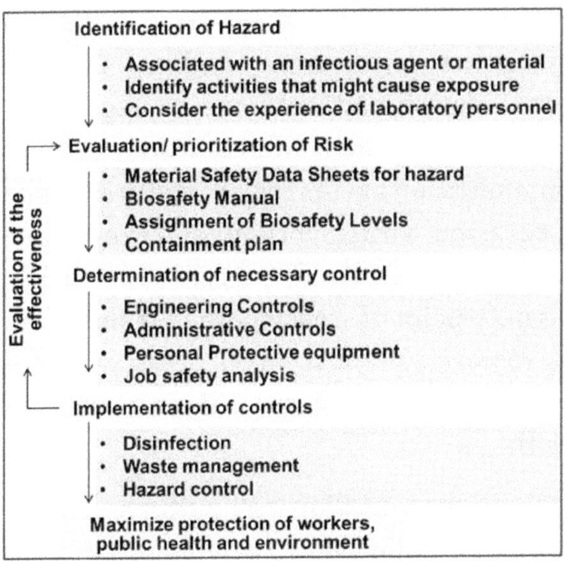

Figure 2. Flow diagram illustrating the summarizing the biosafety assessment and management process while handling cell culture-based experiments. Flow chart is inspired by reference [55].

Understanding and having a complete analysis of the intrinsic infections of cell cultures help to perform well and safe mammalian cell culture. To assess biological risks connected with the mammalian cell cultures, three intrinsic properties related to cell cultures should be considered: the species of origin, the cell type or type of tissue (the organ of origin of the cell line), and the status of the culture. Correspondingly, mammalian cells other than human cells render less risk; still, some infectious agents are proficient at crossing one species to another species, leading to zoonosis. Highly reported infections of viruses comprise hantavirus, hemorrhagic fever viruses, bird Influenza virus, and severe acute respiratory syndrome (SARS) associated virus. Primary cell cultures are created from organ tissues. Highly characterized mammalian cells give the lowest risks compared to primary cultures or less characterized cell lines. Mammalian cells originating from different laboratories without having any proof of identity may cause cross-contamination and pathogen spreading problems, and thereby proper risk assessment and cell characterization are warranted [55, 56]. Several techniques are available for the bio-safety assessment, like RT-PCR, flow cytometry, cytogenetic analysis, DNA fingerprinting, and iso-enzyme analysis. Adventitious contagions of mammalian cell cultures are a vital problem for any activity that involves cell culturing. Contamination agents for cell cultures are bacteria, fungi, mycoplasms, parasites, viruses, prions, and even other animal cells. Modulated experimental results suggest that they spoil the cell cultures. Bio-safety point of view modified mammalian cell cultures for laboratory research, production purposes, or diagnosis purposes they may give support for contaminating materials that cause harm to human health.

5.2 Bioethics and mammalian cell culture

The futuristic technologies in bio-medicine are changing the current concepts and opening up new dimensions. Interestingly as new optimistic channels are opening and expanding, the issues of bioethics are becoming accurate and pertinent. Bioethics is the use of ethical principles in the field of medicine and healthcare. The rational application of ethics in evaluating mammalian cell culture-based experiments is highly warranted, especially during the emerging waves of change in biomedicine. Increased International cross-connection to facilitate open discussion in bioethics and related fields across cross-cultural aspects in bioethics is vital [57]. Several relevant questions arise regarding the private and sensitive use of source data for cells, moral concerns regarding the uses of embryonic and fetal tissue, genetic manipulation, gene therapy, mixing of animal and human cells, tissue banking, legal and intellectual properties associated with ex vivo tissue-engineered cell-based products, an extension of human-ness, etc.

Regarding the humane use of animals, the National Institutes of Health has issued policies as mentioned in the Public Health Service Policy on Humane Care and Use of Laboratory Animals. FDA Human Tissue Task Force and the Center for Biologics Evaluation and Research (CBER) regulates the use of human cells or tissue for implantation, transplantation, infusion, or transfer into a human recipient. The International Society for Stem Cell Research (ISSCR) has also released guidelines for stem cell research and clinical translation. The United States Congress and state legislatures are instrumental in creating laws concerning bioethics. Several professional bioethics organizations, including the American Society for Bioethics and Humanities, American Society for Law, Medicine, and Ethics, Canadian Bioethics Society, provide a platform for discussion over bioethics [57]. Several public institutions supported by academicians and researcher-based initiative for propagating public dialog plays a vital role in educating the masses.

6. SIGNIFICANCE OF IPR ON INDUSTRIAL AND ACADEMIC SCALE

Intellectual property rights (IPR) prevail in any primitive design of the human brain, such as methodical design. IPR mentions the lawful rights agreed to the designer for guarding his innovation for a definite period. These lawful rights grant special rights to the originator or his lender to exploit his idea for a specific period. It is well established that IPR participates in the financial system. It is furthermore overwhelmingly recognized that the intellectualism linked with the originality must be agreed due to value so that products come out of intelligence. The importance of the producer of the technology has turn into lofty and consequently guard the information against unauthorized persons, the use has become a measure, at least sometimes, that would make sure revitalization of the research, investments in developing the technology. IPR helps to look after funds, time, capital, endeavor invested by the producer of an intellectual idea; as a result, IPR, in this way, encourages the profitable encouragement of a realm by encouraging positive competition and heartening trade and industry [58].

The industries have reputations in discussions about IPR strategies, and they are in the face line for controversies about the association among IPRs, R&D incentives, cost, and right to use to supplies [59]. Although, some discussions on the critical issue are relatively little practical proof to support developing IPR policy. This experimental evidence on IP and products inspect practical issues are the primary sources of the data. The industrial sector is composite and much synchronized in the majority of economies. Looking cross-nationally, the contrast among the countries in their perspective on these essential policy affairs generates some additional provocations. In a cosmopolitan industry having control over research and development conveniences in many countries, anticipating a successful transnational technology, goods are raised and developed internationally and are commercialized worldwide. Still, retails are nationalized, with no considerable uniformity across the nations in IPR authorities and various public health care organizations. IPRs may shore up significant discrepancies to price across the nations in returns and demand to prices. These discrepancies in the prices may potentially develop new local and global disagreements. Prominently, for any nation, the essential exchange in IPR regulation options is incredibly dependent on the organizations and function of its health care system.

While having a commendable collaboration, the complete fulfillment of a patent portfolio is to give equal rights for industry and academic institutions. In many countries, research organizations pursuing research in academic institutions, despite their most important work in society as a generator of the intellectual idea, the main concern is to be to deal with IP in a proficient mode. All academic institutions must become accustomed to this development to successfully fulfill the responsibility entrusted to a national or regional innovation ecosystem.

On the supply side, goods safety, supervision of manufacturing, and legal frameworks leading technology transfer among public-funded academic institutions and money-making industries playing an equal role in determining competition. Providing IPR policy to academic institutions has a favorable outcome and various settlements for shareholders. The most significant overarching advantage of these IPR policies was pronounced increases involvement in improving the global innovation performance, i.e., ultimately leading to improving the marketable products and processes. The development of spin-out companies from universities is also growing at a faster rate. The critical part is that the university should own the background IP. Then a resource of external financial support is necessary to finance the start-up company. IPR affairs at academic institutions glow enormous meandering return impending for the national economy. Publishing articles regarding innovations play an essential role in the profession of academic scientists. Participating in knowledge transfer from academia to manufacturing industries can promote academic entrepreneurship. Moreover, these patents have precious information than other publishing articles. Thus, utilizing and increasing patent writing might be beneficial in scientific research. Appropriate IPR policies and tractable technology transfer professionals play a pivotal role in streamlining the necessary work-frame. Published patents improve the economy and reputation of the academic institutions as well as the researchers.

7. CONCISE SUMMARY

The small size and distinctive physicochemical features of nanomaterials have played a key role in advancing the field of biomedicine. Nanomedicine's potential is also dependent on the skillful manipulation of various nanoparticles with adjustable characteristics to regulate their nano-bio interactions for medical purposes. Understanding how nanoparticles interact with biological systems will enable the identification of advantageous physicochemical characteristics that can improve biocompatibility and therapeutic effectiveness without causing any negative consequences. Our understanding of the toxicological effects of manmade nanomaterials is currently incomplete, which hinders the practical application of nanomedicine. The process of modifying the surface of nanoparticles with specific targeting molecules can create highly efficient nanoparticle delivery systems for a wide range of biological applications and tailored therapies. Therefore, utilizing in vitro 2D and 3D cell culture methods helps expedite the assessment of how well nanoparticles interact with living cells and their potential to cause harm, which can ultimately inform the targeted use of nanoparticles in treating certain diseases [60]. Nanoparticles are increasingly employed in a diverse range of cell and tissue-specific scientific investigations, such as cell analysis, in vivo imaging, biosensors, and theranostics. Therefore, the matter of biosafety and bioethics has emerged as a crucial concern when working with mammalian cell cultures. This chapter provides a concise overview of the key elements of biosafety and bioethics that are relevant to investigations with nanomaterials. To summarize, MCC plays a crucial role in contemporary biomedicine and has numerous applications in the detection and treatment of human diseases. Cell culture processes are characterized by their reliability, reproducibility, and lack of bias. However, the process of growing cells might be intricate on certain occasions. The extensive possibilities of utilizing MCC approaches to tackle basic and translational research inquiries have clarified the crucial considerations for establishing a cell culture laboratory. Specifically, 3D organoid culture techniques have been developed to replicate the cellular conditions found in living organisms. Sequencing, mapping, and annotating the genome and its genetic code have become a top concern in biotechnology, particularly in order to comprehend the interaction between nanoparticles and mammalian cells. The process of reporting and categorizing the found gene sequences is crucial for scientific advancement and the development of treatments targeted at certain diseases. The application of nanotechnology in research has made substantial contributions to various scientific disciplines and related sectors. Therefore, the integration of nanotechnology with the mammalian cell culture system can yield a scientific remedy and offer substantial advantages to society as a whole. Therefore, intellectual property rights are crucial for safeguarding the innovator's authority over their discovery. It is crucial to have a strong comprehension of the intellectual property rights (IPR) rules and the protocol for transferring technology. Academic institutions and government entities can collaborate to establish an amicable foundation for effective policy administration. Gaining a more profound comprehension of how nanoparticles interact with cells and developing advanced nanocarriers can usher in a new era of cutting-edge treatments and theranostics.

At a Glance

Modern medicine has challenges in developing a pharmaceutical delivery system that precisely targets specific locations and distributes medications in a controlled manner to avoid opsonization and reduce bio-toxicity. Nanoparticles are materials that possess dimensions at the nanoscale and can originate from either natural or artificial sources. Engineered nanomaterials play a crucial role in the domains of nanomedicine and nanobiotechnology. Furthermore, customized nanoparticles offer therapeutic advantages by enhancing specificity through their unique designed characteristics. Moreover, these nanoparticles that are designed for specific applications have a notable capacity to interact with diverse biological components, including proteins, genetic materials, cell membranes, and organelles, at the interface between nanotechnology and biology. In addition, the surface can be modified with targeting molecules such as small molecule ligands, monoclonal antibodies, aptamers, cell-penetrating peptides, and proteins to enhance the ability of nanoparticles to specifically target certain tissues. This paper presents the latest advancements in therapeutics and theranostics that utilize nanoparticles. A comprehensive understanding of the best methods for preparing nanomaterials, their physical and chemical properties, strategies for modifying their surfaces, and their ability to interact with living organisms can greatly aid in the successful use of nanomaterials in medical contexts. Engineered nanoparticles are in high demand in modern medicine because to their wide range of uses, including bio-imaging, theranostics, tissue engineering, sensors, drug and nucleic acid delivery, and therapeutics. Mammalian cell assays, both in 2D and 3D formats, are frequently used for disease modeling, drug screening, drug development, and toxicity assessment. Researchers have been able to explore a wide range of biologically important issues because to advancements in cell culture technologies and nanotechnology. This chapter explores the properties of nanoparticles, different methods of targeted delivery, biological assessment, and theranostics. This chapter focuses on the topics of biosafety, bioethics, and intellectual property rights in relation to mammalian cell culture. It examines these subjects from both industrial and academic perspectives.

ABBREVIATIONS

CBER	Centre for Biologics Evaluation and Research
CT	Computed Tomography
EGFR	Epidermal Growth Factor Receptor
EPR	Enhanced Permeability and Retention
FGFR	Fibroblast Growth Factor Receptor
HUVEC	Human Umbilical Vein Endothelial Cells
IgM	Immunoglobulin M
IPR	Intellectual Property Rights
ISSCR	International Society for Stem Cell Research
LAIs	Laboratory-Acquired Infections
MMPs	Matrix Metalloproteases

MPI	Magnetic Particle imaging
MRI	Magnetic Resonance Imaging
NIH	National Institutes of Health
NMs	Nanomaterials
NPs	Nanoparticles
NRIF	Near-infrared Fluorescence
NT	Nanotechnology
OCT	Octreotide
PEG	Polyethylene glycol
PET	Positron Emission Tomography
QD	Quantum Dots
SARS	Severe Acute Respiratory Syndrome
SPECT	Single-photon Emission Computed Tomography
SRA	Sequence Read Archive
SWIR	Shortwave Infrared
TME	Tumor Microenvironment
US	Ultrasonography
USFDA	United States Food and Drug Administration
VCAM-1	Vascular Cell Adhesion Molecules-1
VEGFR	Vascular Endothelial Growth Factor Receptors

REFERENCES

1. Poon W, Kingston BR, Ouyang B, Ngo W, Chan WCW. A framework for designing delivery systems. Nat Nanotechnol 2020;15(10):819-829.[10.1038/s41565-020-0759-5]

2. Golombek SK, May JN, Theek B, et al. Tumor targeting via EPR: Strategies to enhance patient responses. Adv Drug Deliv Rev 2018;130:17-38.[10.1016/j.addr.2018.07.007]

3. Mukherjee S, Madamsetty VS, Bhattacharya D, Roy Chowdhury S, Paul MK, Mukherjee A. Recent Advancements of Nanomedicine in Neurodegenerative Disorders Theranostics. Advanced Functional Materials 2020;30(35).[10.1002/adfm.202003054]

4. Freshney RI. Culture of Animal Cells.[10.1002/9780470649367]

5. Stacey G. Current Developments in Cell Culture Technology. In: New Technologies for Toxicity Testing. 2012;1-13. [10.1007/978-1-4614-3055-1_1]

6. Wilkinson DC, Alva-Ornelas JA, Sucre JMS, et al. Development of a Three-Dimensional Bioengineering Technology to Generate Lung Tissue for Personalized Disease Modeling. STEM CELLS Translational Medicine 2017;6(2):622-633.[10.5966/sctm.2016-0192]

7. Sanna V, Sechi M. Therapeutic Potential of Targeted Nanoparticles and Perspective on Nanotherapies. ACS Medicinal Chemistry Letters 2020;11(6):1069-1073.[10.1021/acsmedchemlett.0c00075]

8. Mitchell MJ, Billingsley MM, Haley RM, Wechsler ME, Peppas NA, Langer R. Engineering precision nanoparticles for drug delivery. Nature Reviews Drug Discovery 2020;20(2):101-124. [10.1038/s41573-020-0090-8]

9. Li X, Wang L, Fan Y, Feng Q, Cui F-z. Biocompatibility and Toxicity of Nanoparticles and Nanotubes. Journal of Nanomaterials 2012;2012:1-19.[10.1155/2012/548389]

10. Patra JK, Das G, Fraceto LF, et al. Nano based drug delivery systems: recent developments and future prospects. Journal of Nanobiotechnology 2018;16(1).[10.1186/s12951-018-0392-8]

11. Wagner AM, Knipe JM, Orive G, Peppas NA. Quantum dots in biomedical applications. Acta Biomaterialia 2019;94:44-63. [10.1016/j.actbio.2019.05.022]

12. Beltrán-Gracia E, López-Camacho A, Higuera-Ciapara I, Velázquez-Fernández JB, Vallejo-Cardona AA. Nanomedicine review: clinical developments in liposomal applications. Cancer Nanotechnology 2019;10(1).[10.1186/s12645-019-0055-y]

13. Bulbake U, Doppalapudi S, Kommineni N, Khan W. Liposomal Formulations in Clinical Use: An Updated Review. Pharmaceutics 2017;9(4).[10.3390/pharmaceutics9020012]

14. Liu C, Zhang L, Zhu W, et al. Barriers and Strategies of Cationic Liposomes for Cancer Gene Therapy. Molecular Therapy - Methods & Clinical Development 2020;18:751-764.[10.1016/j.omtm.2020.07.015]

15. Jeevanandam J, Barhoum A, Chan YS, Dufresne A, Danquah MK. Review on nanoparticles and nanostructured materials: history, sources, toxicity and regulations. Beilstein Journal of Nanotechnology 2018;9:1050-1074.[10.3762/bjnano.9.98]

16. Zein R, Sharrouf W, Selting K. Physical Properties of Nanoparticles That Result in Improved Cancer Targeting. Journal of Oncology 2020;2020:1-16.[10.1155/2020/5194780]

17. Blanco E, Shen H, Ferrari M. Principles of nanoparticle design for overcoming biological barriers to drug delivery. Nature Biotechnology 2015;33(9):941-951.[10.1038/nbt.3330]

18. Huang Y-W, Cambre M, Lee H-J. The Toxicity of Nanoparticles Depends on Multiple Molecular and Physicochemical Mechanisms. International Journal of Molecular Sciences 2017;18(12).[10.3390/ijms18122702]

19. Sanità G, Carrese B, Lamberti A. Nanoparticle Surface Functionalization: How to Improve Biocompatibility and Cellular Internalization. Frontiers in Molecular Biosciences 2020;7.[10.3389/fmolb.2020.587012]

20. Gonda A, Zhao N, Shah JV, et al. Engineering Tumor-Targeting Nanoparticles as Vehicles for Precision Nanomedicine. Med One 2019;4.[10.20900/mo.20190021]

21. Rawat M, Yadukrishnan P, Kumar N. Mechanisms of Action of Nanoparticles in Living Systems. In: Microbial Biotechnology

in Environmental Monitoring and Cleanup. 2018;220-236. [10.4018/978-1-5225-3126-5.ch014]

22. Abbas Z, Rehman S. An Overview of Cancer Treatment Modalities. In: Neoplasm. 2018.[10.5772/intechopen.76558]

23. Chenthamara D, Subramaniam S, Ramakrishnan SG, et al. Therapeutic efficacy of nanoparticles and routes of administration. Biomaterials Research 2019;23(1).[10.1186/s40824-019-0166-x]

24. Kim ST, Saha K, Kim C, Rotello VM. The Role of Surface Functionality in Determining Nanoparticle Cytotoxicity. Accounts of Chemical Research 2013;46(3):681-691.[10.1021/ar3000647]

25. Albanese A, Tang PS, Chan WCW. The Effect of Nanoparticle Size, Shape, and Surface Chemistry on Biological Systems. Annual Review of Biomedical Engineering 2012;14(1):1-16.[10.1146/annurev-bioeng-071811-150124]

26. Salmaso S, Caliceti P. Stealth Properties to Improve Therapeutic Efficacy of Drug Nanocarriers. Journal of Drug Delivery 2013;2013:1-19.[10.1155/2013/374252]

27. Gal N, Charwat V, Städler B, Reimhult E. Poly(ethylene glycol) Grafting of Nanoparticles Prevents Uptake by Cells and Transport Through Cell Barrier Layers Regardless of Shear Flow and Particle Size. ACS Biomaterials Science & Engineering 2019;5(9):4355-4365.[10.1021/acsbiomaterials.9b00611]

28. Large DE, Soucy JR, Hebert J, Auguste DT. Advances in Receptor-Mediated, Tumor-Targeted Drug Delivery. Advanced Therapeutics 2019;2(1).[10.1002/adtp.201800091]

29. Mukherjee A, Madamsetty VS, Paul MK, Mukherjee S. Recent Advancements of Nanomedicine towards Antiangiogenic Therapy in Cancer. International Journal of Molecular Sciences 2020;21(2).[10.3390/ijms21020455]

30. Frigerio B, Bizzoni C, Jansen G, et al. Folate receptors and transporters: biological role and diagnostic/therapeutic

targets in cancer and other diseases. Journal of Experimental & Clinical Cancer Research 2019;38(1).[10.1186/s13046-019-1123-1]

31. Siwak DR, Carey M, Hennessy BT, et al. Targeting the Epidermal Growth Factor Receptor in Epithelial Ovarian Cancer: Current Knowledge and Future Challenges. Journal of Oncology 2010;2010:1-20.[10.1155/2010/568938]

32. Terada T, Mizobata M, Kawakami S, Yamashita F, Hashida M. Optimization of tumor-selective targeting by basic fibroblast growth factor-binding peptide grafted PEGylated liposomes. Journal of Controlled Release 2007;119(3):262-270.[10.1016/j.jconrel.2007.01.018]

33. Holzmann, Marian. Importance of Translational Research for Targeting Fibroblast Growth Factor Receptor Signaling in Cancer. Cells 2019;8(10).[10.3390/cells8101191]

34. Mukherjee A, Waters AK, Babic I, et al. Antibody drug conjugates: Progress, pitfalls, and promises. Human Antibodies 2018;27(1):53-62.[10.3233/hab-180348]

35. Roma-Rodrigues C, Mendes R, Baptista P, Fernandes A. Targeting Tumor Microenvironment for Cancer Therapy. International Journal of Molecular Sciences 2019;20(4).[10.3390/ijms20040840]

36. Sakurai, Akita, Harashima. Targeting Tumor Endothelial Cells with Nanoparticles. International Journal of Molecular Sciences 2019;20(23).[10.3390/ijms20235819]

37. Silva S, Almeida A, Vale N. Combination of Cell-Penetrating Peptides with Nanoparticles for Therapeutic Application: A Review. Biomolecules 2019;9(1).[10.3390/biom9010022]

38. Kapp TG, Rechenmacher F, Neubauer S, et al. A Comprehensive Evaluation of the Activity and Selectivity Profile of Ligands for RGD-binding Integrins. Scientific Reports 2017;7(1).[10.1038/srep39805]

39. Maghsoudnia N, Eftekhari RB, Sohi AN, Zamzami A, Dorkoosh FA. Application of nano-based systems for drug delivery and targeting: a review. Journal of Nanoparticle Research 2020;22(8). [10.1007/s11051-020-04959-8]

40. Issa B, M. Obaidat I. Magnetic Nanoparticles as MRI Contrast Agents. In: Magnetic Resonance Imaging. 2019. [10.5772/intechopen.84649]

41. Colino C, Millán C, Lanao J. Nanoparticles for Signaling in Biodiagnosis and Treatment of Infectious Diseases. International Journal of Molecular Sciences 2018;19(6).[10.3390/ijms19061627]

42. Zhang R, Belwal T, Li L, Lin X, Xu Y, Luo Z. Nanomaterial-based biosensors for sensing key foodborne pathogens: Advances from recent decades. Comprehensive Reviews in Food Science and Food Safety 2020;19(4):1465-1487.[10.1111/1541-4337.12576]

43. Lin Q, Fathi P, Chen X. Nanoparticle delivery in vivo: A fresh look from intravital imaging. EBioMedicine 2020;59.[10.1016/j.ebiom.2020.102958]

44. Su Y, Xie Z, Kim GB, Dong C, Yang J. Design Strategies and Applications of Circulating Cell-Mediated Drug Delivery Systems. ACS Biomaterials Science & Engineering 2015;1(4):201-217.[10.1021/ab500179h]

45. Wolfbeis OS. An overview of nanoparticles commonly used in fluorescent bioimaging. Chemical Society Reviews 2015;44(14):4743-4768.[10.1039/c4cs00392f]

46. Van der Meel R, Sulheim E, Shi Y, Kiessling F, Mulder WJM, Lammers T. Smart cancer nanomedicine. Nature Nanotechnology 2019;14(11):1007-1017.[10.1038/s41565-019-0567-y]

47. Benito-Alifonso D, Richichi B, Baldoneschi V, et al. Quantum Dot-Based Probes for Labeling and Imaging of Cells that Express Matrix Metalloproteinases. ACS Omega2018;3(8):9822-9826.[10.1021/acsomega.8b00633]

48. Siemer S, Wünsch D, Khamis A, et al. Nano Meets Micro-Translational Nanotechnology in Medicine: Nano-Based Applications for Early Tumor Detection and Therapy. Nanomaterials 2020;10(2).[10.3390/nano10020383]

49. Che Y, Feng S, Guo J, et al. In vivo live imaging of bone using shortwave infrared fluorescent quantum dots. Nanoscale 2020;12(43):22022-22029.[10.1039/d0nr06261h]

50. Mukherjee A, Paul M, Mukherjee S. Recent Progress in the Theranostics Application of Nanomedicine in Lung Cancer. Cancers 2019;11(5).[10.3390/cancers11050597]

51. Madamsetty VS, Paul MK, Mukherjee A, Mukherjee S. Functionalization of Nanomaterials and Their Application in Melanoma Cancer Theranostics. ACS Biomaterials Science & Engineering 2019;6(1):167-181.[10.1021/acsbiomaterials.9b01426]

52. Agrawal V, Paul MK, Mukhopadhyay AK. 6-Mercaptopurine and Daunorubicin Double Drug Liposomes—Preparation, Drug-Drug Interaction and Characterization. Journal of Liposome Research 2008;15(3-4):141-155.[10.1080/08982100500364081]

53. Chen D-W, Cheng L, Huang F, et al. GE11-modified liposomes for non-small cell lung cancer targeting: preparation, ex vitro and in vivo evaluation. International Journal of Nanomedicine 2014;10.2147/ijn.S53310.[10.2147/ijn.S53310]

54. Song X-l, Ju R-j, Xiao Y, et al. Application of multifunctional targeting epirubicin liposomes in the treatment of non-small-cell lung cancer. International Journal of Nanomedicine 2017;12:7433-7451.[10.2147/ijn.S141787]

55. Herman P, Pauwels K. Biosafety Recommendations on the Handling of Animal Cell Cultures. In: Animal Cell Culture. 2015:689-716.[10.1007/978-3-319-10320-4_22]

56. Geraghty RJ, Capes-Davis A, Davis JM, et al. Guidelines for the use of cell lines in biomedical research. British Journal of Cancer 2014;111(6):1021-1046.[10.1038/bjc.2014.166]

57. Segers S, Mertes H, de Wert G, Dondorp W, Pennings G. Balancing Ethical Pros and Cons of Stem Cell Derived Gametes. Annals of Biomedical Engineering 2017;45(7):1620-1632.[10.1007/s10439-017-1793-9]

58. Prabu SL, Suriyaprakash TNK, Thirumurugan R. Introductory Chapter: Intellectual Property Rights. In: Intellectual Property Rights. 2017. [10.5772/intechopen.69359]

59. Lemley MA. Intellectual property in the new technological age :2016. Berkeley, CA: Clause 8 Pub.; 2016

60. Abdel Fattah AR, Ranga A. Nanoparticles as Versatile Tools for Mechanotransduction in Tissues and Organoids. Frontiers in Bioengineering and Biotechnology 2020;8.[10.3389/fbioe.2020.00240]

INDEX

www.ingramcontent.com/pod-product-compliance
Ingram Content Group UK Ltd.
Pitfield, Milton Keynes, MK11 3LW, UK
UKHW020048200726
13845UKWH00045B/61

9 789349 964778